PERSPECTIVE AND PROJECTIVE GEOMETRY

Published by Princeton University Press
41 William Street, Princeton, New Jersey 08540
6 Oxford Street, Woodstock, Oxfordshire OX20 1TR

press.princeton.edu

Library of Congress Control Number: 2019937983
ISBN 978-0-691-19655-8
ISBN (pbk) 978-0-691-19656-5
ISBN (e-bk) 978-0-691-19738-8

British Library Cataloging-in-Publication Data is available

Editorial: Vickie Kearn, Susannah Shoemaker and Lauren Bucca
Production Editorial: Jenny Wolkowicki
Text design: Pamela Schnitter
Cover design: Pamela Schnitter
Production: Jacqueline Poirier
Publicity: Matthew Taylor and Katie Lewis
Copyeditor: Bhisham Bherwani

Cover art: Illustration of an imaginary art museum featuring a sketch and
sculptures inspired by the works of mathematician Brook Taylor and artists
Donald Judd and Roger Jorgensen. Black and white charcoal on gray paper.
By Fumiko Futamura

This book has been composed in Adobe Text Pro and Helvetica LT Std

Printed on acid-free paper. ∞

Printed in the United States of America

10 9 8 7 6 5 4 3 2 1

Contents

A comment on page numbering: almost every module begins with a one-page picture that is also a math/art puzzle meant to be removed for ease of drawing. The in-class worksheet follows immediately after.

0   Introduction and First Action   1

1   Window Taping: The After Math   9

     Appendix: A Working Definition of *n*-Point Perspective   22

2   Drawing ART   25

3   What's the Image of a Line?   33

4   The Geometry of $\mathbb{R}^2$ and $\mathbb{R}^3$ 43

     4.1   Euclidean Geometry: A Point of Comparison   43

     4.2   Euclidean Geometry Revisited: Similarities and Invariants   51

5   Extended Euclidean Space: To Infinity and Beyond   63

6   Of Meshes and Maps   75

     6.1   Field Trip: Perspective Poster   87

7   Desargues's Theorem   91

     7.1   Exploration and Discovery   93

     7.2   Working toward a Proof   103

8   Collineations   117

     8.1   How Projective Geometry Functions   119

     8.2   Reflecting on Homologies and Harmonic Sets   131

     8.3   Elations (or, How to Be Transported in a Math Class)   141

9 Dynamic Cubes and Viewing Distance 145

10 Drawing Boxes and Cubes in Two-Point Perspective 157

11 Perspective by the Numbers 171

    11.1 Discovering the Cross Ratio 173

    11.2 Eves's Theorem 189

    11.3 An Angle on Perspective: Casey's Theorem 203

12 Coordinate Geometry 211

    12.1 Euclidean Geometry Enhanced with Algebra 213

    12.2 Introduction to Homogeneous Coordinates 219

13 The Shape of Extended Space 225

Appendix G Introduction to GeoGebra 235

Appendix R Reference Manual 245

Appendix W Writing Mathematical Prose 259

    W.1 Getting Started 261

        W.1.1 *Why* We Write Proofs, and What That Means
             for *How* You Write Proofs 261

        W.1.2 Mechanics and Conventions 261

    W.2 Pronouns and Active Voice 263

    W.3 Introducing and Using Variables, Constants,
        and Other Mathematical Symbols 265

    W.4 Punctuation with Algebraic Expressions in the Sentence 267

    W.5 Paragraphs and Lines 269

W.6 Figures 271

     W.6.1 Formatting the Figure 271

     W.6.2 Referring to Figures 271

Acknowledgments 273

Bibliography 275

Index 279

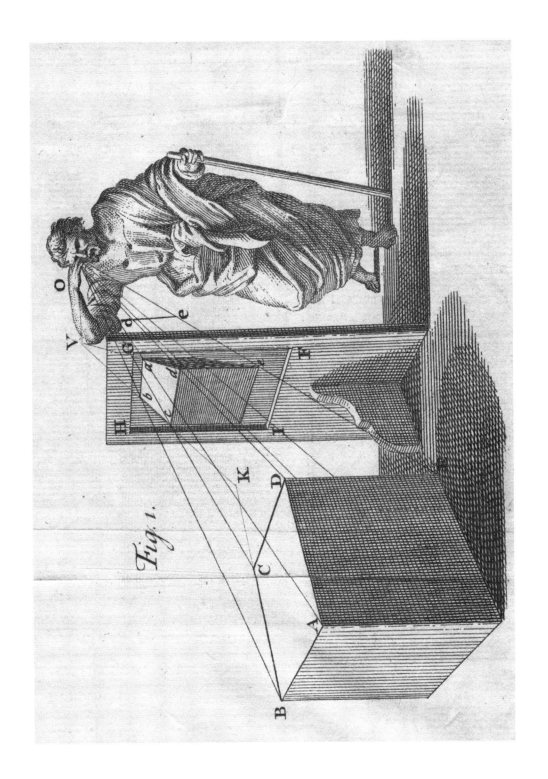

FIGURE 0: Looking at the world through a window. [For use with the INTRODUCTION AND FIRST ACTION module.]
Courtesy of the Max Planck Institute for the History of Science, Berlin

0

Introduction and First Action

Looking at the World through the Window of Mathematics

Perspective and Projective Geometry is a course that will change the way that you look at the world, and we mean that literally.

In this course, you will take photographs, you will analyze perspective pictures and draw pictures of your own, and you will explore the geometry that explains how we fit a three-dimensional world onto a two-dimensional canvas. Along the way, you will get invaluable practice with making logical arguments (that is, writing mathematical proofs of statements that are true and refuting statements that are false). By combining art and geometry, we are following in a tradition that builds on centuries of exploration. Indeed, the very first treatise on perspective art—Leon Battista Alberti's *Della Pittura* (On Painting)—includes this exhortation from the author:

> It would please me if the painter were as learned as possible in all the liberal arts, but first of all I desire that he know geometry. ... Our instruction in which all the perfect absolute art of painting is explained will be easily understood by the geometrician, but one who is ignorant in geometry will not understand these or any other rules of painting. Therefore, I assert that it is necessary for the painter to learn geometry. [3, p. 90]

Knowing how to look at the world is more than just a fancy, aesthetic luxury. The geometry that helped Renaissance artists create breathtaking, realistic images five or six centuries ago also helped those same societies construct maps that allowed them to navigate across the globe; it helped them understand the emerging warfare of ballistic cannons; and it helped them build fortresses that could withstand ballistic attacks.

Indeed, Girard Desargues—the author of one of the central theorems of this book—extolled the subject of geometry because of its usefulness not only in the world at large, but also to his own well-being:

> I freely confess that I never had taste for study or research either in physics or geometry except in so far as they could serve as a means of arriving at some sort of knowledge of the proximate causes for the good and convenience of life, in maintaining health, in the practice of some art ... having observed that a good part of the arts is based on geometry ... that of perspective in particular [33].

That very same geometry that defined an age named for "new birth" is making its own new strides in our modern world. During our lifetime, video games have moved from the two-dimensional mazes of *Pac-Man* into the immersive full-body experiences. Your parents and grandparents watched the flat worlds of the Flintstones and the Simpsons, but as the calendar ticked over into the most recent millennium, movies like *Shrek* and *Frozen* started bringing pixelated characters to life. And just as in the Renaissance, the uses of this projective geometry have spilled out over the edges of art into many areas of practical technology, paving the way for unprecedented progress in medical imaging, in geological exploration, and in 360-degree map views that have made commercial successes of applications like Zillow and Google. Knowing how to see the world is powerful, and this course will help you to harness that power.

We learn by doing, and so each lesson uses the following format (with occasional, minor variations). We begin each lesson with a picture puzzle. This puzzle comes with a module that has questions and occasional definitions that will help you and your fellow classmates construct an understanding of the geometry that allows us to solve that puzzle and others like it. At the end of the module, you will see three kinds of homework questions:

- short answer exercises (denoted by a Ⓔ symbol),
- "art" exercises (denoted by a △ symbol) that ask you to create drawings or photographs with certain properties, and
- proof or counterexample questions (denoted by a ⊡ symbol) that build your reasoning and exposition skills.

At the end of this book, we include the key definitions and theorems in a Reference Manual to aid you in reviewing and studying.

Even before you begin the first module, we hope you will have the experience of looking at the world and projecting an image of it onto a two-dimensional canvas. In particular, you and your classmates will get to draw pictures on windows.

A Window into Perspective

The word *perspective* comes from the Medieval Latin roots *per* ("through") and *specere* ("look at"—the same root that gives us "spectacles"). So perspective art literally intends for us to look through a window to portray the objects that lie on the other side. As Alberti instructed aspiring painters, "When [artists] fill the circumscribed places with colors, they should only seek to present the forms of things seen on this plane as if it were of transparent glass." Alberti's book had a huge influence on numerous scholars and artists of his time, including Leonardo da Vinci, Piero della Francesca, Albrecht Dürer, and Gerard Desargues. Three centuries after Alberti's treatise appeared, the mathematician Brook Taylor (of Taylor series fame) illustrated exactly such a *through-the-window* projection (his Figure 1, our Figure 0) in the preface of his book, *New Principles of Linear Perspective* published in 1719 [51].

The description of "looking through a window" wasn't merely a metaphor, and it wasn't meant as a mere illustration for descriptions that appeared in a book. Artists throughout the ages have practiced the actual physical act of drawing the world on a window they gazed through. When Leonardo da Vinci instructed painters on "how to draw a site correctly" [36, p. 65], he wrote,

> Have a sheet of glass as large as a half-sheet of royal folio paper, and place it firmly in front of your eyes; that is between the eye and the thing that you draw. Then place yourself at a distance of two-thirds of a braccio [arm's length] from your eye to the

glass, and hold your head with an instrument in such a way that you cannot move your head in the least. Then close or cover one eye, and with a brush or with a pencil of red chalk draw on the glass what appears beyond it.

Even before you start drawing images on paper, therefore, you ought to experience drawing through a window so that you can better understand some of the implications of projecting our three-dimensional world in this way.

The following set of instructions leads you through such an exercise.

Instructions for Window Taping

1. Get into a group of three or four people and choose one person to be the *Art Director*. The others will be *Artists* (and *Holders of Windows*, if using plexiglass).

   (a) *Art Director*: Stand or sit at a fixed location, close one eye, and look through the window with the other (see Figure 1). Direct the Artists to tape the outline and the most important and instructive features you see on the other side of the window.

   As Leonardo noted above, you will need to "hold your head ... in such a way that you cannot move your head in the least"! In particular, keep your eye fixed in one location. When you start working on a new line, make sure the drafting tape that is already on the window correctly lines up with the features you've already worked on.

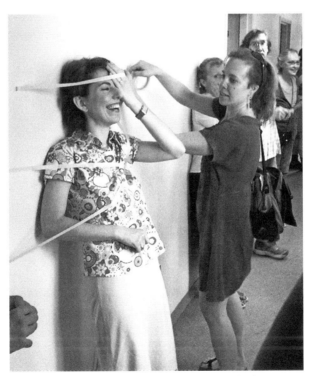

FIGURE 1: Art Director: Although you probably don't need to be taped to the wall, it's still very important to keep your eye fixed in one location!

FIGURE 2: Artists: Use the drafting tape and pay close attention to the Art Director's directions.

FIGURE 3: If you have a camera, take a photograph.

    (b) *Artists*: Paying close attention to the Art Director's directions, place the drafting tape on the windows. It may help to have one person hold one end of the tape and the other person hold the other end; get the directions of lines correct first, and then break the tape to the correct length. At first this will be extremely difficult, because you can't see what the Art Director sees. The job might get easier as you place more tape on the window to use as guidelines.

    (c) (*Holders of Windows*: hold the plexiglass window as still as possible!)

2. When the picture is "done" (or as done as possible for this session), you might want to take a photograph of the finished product from various places, including the

FIGURE 4: If we extend some of the tape lines by adding more tape, we might realize they seem to intersect in a common point.

point of view of the Art Director. If you like, print out your photograph on copier paper and bring it to class.

3. If there are other groups working on similar drawings, you might also wander around and look at the other groups' pictures. Try to see if you can stand or sit where the Art Director was, to see the drawing from the Art Director's point of view.

4. You will notice that some parallel lines in the real world (such as, probably, the vertical lines) have corresponding tape images that are likewise vertical. But some sets of parallel lines in the real world have tape images that seem to tilt. If you extend these lines by adding more tape, you might notice that all of these lines intersect in one spot on the window (see Figure 4). Understanding why this intersection happens, and what the geometric significance of this intersection point is, will be the topic of the first module.

5. Before you come back to the classroom, clean up the tape! Drafting tape usually leaves no residue on windows, especially if you remove it promptly.

Questions for Review

1. Which job was harder, the Art Director's, or the Artists'? Why?
2. Why did the Art Director need to cover one eye?
3. Why was it hard for the Artists to figure out where to put the tape?
4. What does it mean for a line in the real world to be parallel to the window?
5. If the lines in the real world are parallel to the window, what did you notice about their taped images?
6. One collection of lines met at a point. What did this point have to do with the Art Director?

7. What is the relationship between the window and the lines in the real world whose taped images are the lines that met at a point?
8. If you looked at other groups' pictures, could you figure out where the Art Director had been standing or sitting?

About the cover: The cover of this book depicts a room in an imaginary art museum that illustrates, literally, many important geometric concepts that will arise in this book. On the wall hangs a reproduction of Figure 0; the geometry of looking through a window will appear in Chapter 1, and the surprisingly tricky problem of drawing cubes is the subject of Chapters 9 and 10.

The regularly repeated cubes on the floor are inspired by the minimalist artwork of Donald Judd, hinting at the concepts of geometric division, cross ratio, and projective collineations found in chapters 6, 11, and 8, respectively.

The sculpture with the triangular opening was inspired by the work of Roger Jorgensen; the triangular pools of light cast by the sculpture depict a consequence of Desargues's theorem, which the reader will prove in Chapter 7. The space invites you to explore these and many other ideas found in the rooms beyond.

The work, by Fumiko Futamura, was planned in GeoGebra, then geometrically constructed and hand-drawn in black and white charcoal on gray paper.

FIGURE 1.0: Yuxun Sun (F&M Class of 2014) looks out of a window that he and his classmates have taped. See exercises below. [For use with the WINDOW TAPING: THE AFTER MATH module.] With permission of Yuxun Sun

5. [T/F] If a plane in $\mathbb{R}^3$ and a line not lying in that plane are parallel, then they do not intersect.

6. [T/F] If a line and a plane, both in $\mathbb{R}^3$, do not intersect, then they are parallel.

7. [T/F] If the line ℓ is parallel to the plane ω, then there is some line $k \subset \omega$ that is parallel to ℓ.

8. [T/F] If there is some line $k \subset \omega$ that is parallel to a line ℓ, then ℓ is parallel to ω.

(Note: *Unless we say otherwise, we use the standard geometric convention of naming points with italicized capital letters, lines with italicized lowercase letters, and planes with lowercase Greek letters.*)

Real-World Lines and their Images

For the following questions, refer either to a photograph you took of your window-taping exercise, or to Figure 1.0. We will be looking at lines in the real world and at their images on the window (or even more specifically, the images of the tape in the photograph).

9. In the photograph, identify the images of a set of three lines that are parallel to each other in the real world and also parallel to the picture plane (the window). Use your straightedge to extend the line segments to the edges of the paper. Are these images in the photograph parallel?

10. Now identify the images of a set of three lines that are parallel to each other in the real world but **not** parallel to the picture plane. Use your straightedge to extend these line segments. Are these images in your photograph parallel?

11. In general, we expect that a pair of lines that are not parallel will intersect in a single point, but we might be surprised if a set of *three* lines that are not parallel happen to intersect in a single point. Consider the lines from question 10. Describe how the intersection point you located in that question is related to the plane of the window, to the original lines, and to the Art Director.

12. Formulate a conjecture:
    If a set of lines in $\mathbb{R}^3$ are parallel to each other and also parallel to the picture plane, then their images _____.
    If a set of lines in $\mathbb{R}^3$ are parallel to each other but **not** parallel to the picture plane, then their images _____.

Side Views and Top Views

In this class, we will often analyze an image by using a *top view* and a *side view* as in Figure 1.1. In this diagram, we see a *viewer*, the *picture plane* (which is drawn edge-on as a line in this diagram), and the *object*. In these two views, the object is a beach towel lying on the horizontal ground.

13. In the side view, the dot labeled A' on the picture plane is the *image* of either the closest corner or the far corner of the towel. Which corner is it? And how do you know? Label that corner of the towel A.

14. On the side-view, label the other corner (the far or the near corner, whichever you didn't choose for question 13) of the towel as B. Use a straightedge to determine B', and then shade in the image of the towel.

TOP VIEWS: LOOKING THROUGH THE NOSE

You discovered on the first day that looking through only one eye makes a huge difference. Rather than try to remember which eye our viewer looks through, when we draw top views, we'll assume that our viewers look through their **noses**. (The assumption is silly, but our diagrams will be *much* easier to draw!)

15. On the top view in Figure 1.1, label the corners A and B of the towel. Draw the four light rays connecting the four corners of the towel to the viewer's nose, and then draw the image of the towel. Label A' and B' on the picture plane.

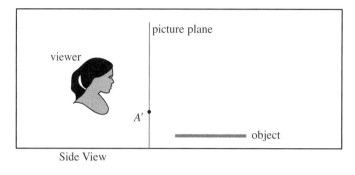

Side View

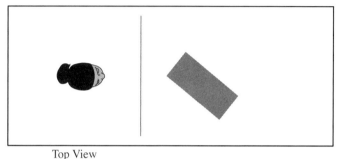

Top View

FIGURE 1.1: Two views of an artist looking at a towel through a picture plane.

Images of Lines

What we have done so far is to look at points and their images. What can we say about lines and their images?

In Figure 1.2, we show a side view of the towel lying on the pebble-strewn ground. We will take advantage of this seemingly obvious fact: *If you are looking at something, you can see it. If you are not looking at something, you can't see it.* In particular, if our viewer looks up at the sky (as we indicate in the side view), she does not see the ground. (When we talk about looking in one direction, it helps to think of a "line of sight," as though the viewer is looking through a straw and has no peripheral vision).

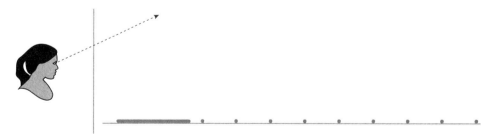

FIGURE 1.2: Side view showing a viewer facing a towel on the pebble-strewn ground. When the viewer looks up, she does not see the ground so she does not draw it.

16. In Figure 1.2, use a straightedge to draw the light rays connecting the viewer's eye to the pebbles.

(a) As the pebbles get further and further away from the viewer, what can you say about the light rays?

(b) Draw the images of the pebbles on the picture plane. As the pebbles get further from the viewer, what can you say about their images?

(c) Suppose the line of pebbles goes on forever. Locate the point on the picture plane where the viewer goes from *seeing* the pebbles to *not seeing* the pebbles. Because the ground appears to vanish at this point on the picture plane, we call this point the *vanishing point* of the line of pebbles.

(d) Draw the line of sight from the viewer to the vanishing point. How does this line relate to the line of pebbles?

(e) Fill in the blanks to complete the definition:
Given a line ℓ not parallel to the picture plane ω, the *vanishing point* V_ℓ of the line is the intersection of the plane _____ with the line through _____ that is parallel to _____.

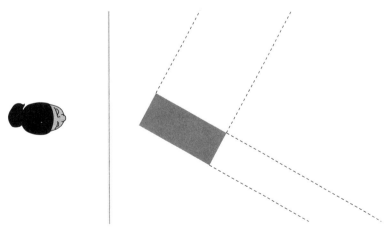

FIGURE 1.3: Top view showing a viewer looking at a towel, with the edges of the towel extended.

17. Let us repeat this procedure with the four lines in Figure 1.3. Add some pebbles to the lines that extend the edge of the towel, and then draw their images.

(a) Locate the vanishing points of these four lines. How many vanishing points are there?

(b) Draw the line(s) that connect(s) the viewer's nose to the vanishing point(s). What can you say about how the line(s) relate(s) to the edges of the towel?

18. In the space below, draw your own side view and top view, showing a viewer, the vertical picture plane, and two horizontal lines that are parallel to the picture plane. Answer these questions about your side and top views:
(a) Must the two lines you draw be parallel to one another?
(b) Locate the image of each line in the side view.
(c) Locate the image of each line in the top view.
(d) How does the image of each line relate to the original line?
(e) How many vanishing points do each of these lines have?

CHECKING THE CONJECTURE

19. Compare your conjecture from question 12 to your answers to questions 16–18. Do your answers agree with one another?

20. In the drawing below, complete the sketch of the towel in perspective (that is, as it should appear in the picture plane), using your answers from above. Draw the edges of the towel as solid lines, and the extension of those edges as dashed lines. The point A' is already labeled; label point B' to correspond with your answers to questions 13–15.

21. Why don't points A and B appear in the picture you draw for question 20?

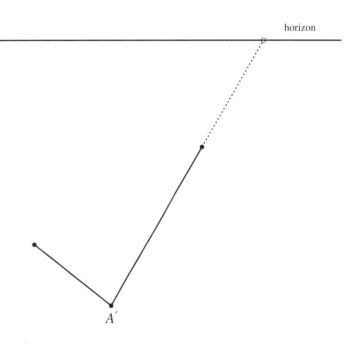

horizon

A'

22. Complete the following set of statements.

     If a line is not parallel to the picture plane, then it has a vanishing point. That vanishing point is located at the intersection of the _____ plane and the line through the artist's eye that is _____.

    A collection of lines that are parallel to each other and also parallel to the picture plane has exactly _____ vanishing point(s).

    A collection of lines that are parallel to each other but not parallel to the picture plane has exactly _____ vanishing point(s).

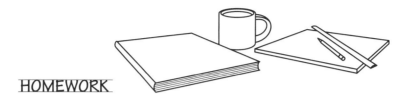

HOMEWORK

EXERCISES

ⓔ 1.1. Consider Figure 1.4, which shows two sketches of hallways. For each sketch, determine whether the artist was standing, sitting, or lying down, and explain how you know.

FIGURE 1.4: Two sketches of hallways for homework exercise ⓔ1.1 (sketches reproduced with permission of Anthony Nocket and Kevin Toenboehm).
Courtesy of Anthony Nocket and Kevin Toenboehm

ⓔ 1.2. Suppose an artist is standing at a distance d from a picture plane, looking at a line segment that is parallel to the picture plane. If the length of the segment is L, can we determine the length l of its image? What other information might we need? Why do we need to know that the segment is parallel to the picture plane?

ⓔ 1.3. In all that we have done so far, we have assumed we look with only one eye. In fact, throughout this course, we will assume that we project images from a single point. But looking with only one eye is not the way that most people actually see the world.

We believe that objects that are further away from us appear to be smaller. But Leonardo da Vinci in his *Codex Urbinas* described why this is not always true when we observe the world with both eyes open. Recreate the argument that he presented in Figure 1.5. That is, explain why Leonardo writes that if we observe objects m and n, both smaller than the distance between our two eyes a and b, the near object "cannot conceal [the further object] entirely." However, if we observe same-sized objects f and r with just one eye s, then "the body f will cover r ... hence the second body of equal size is never seen" [36].

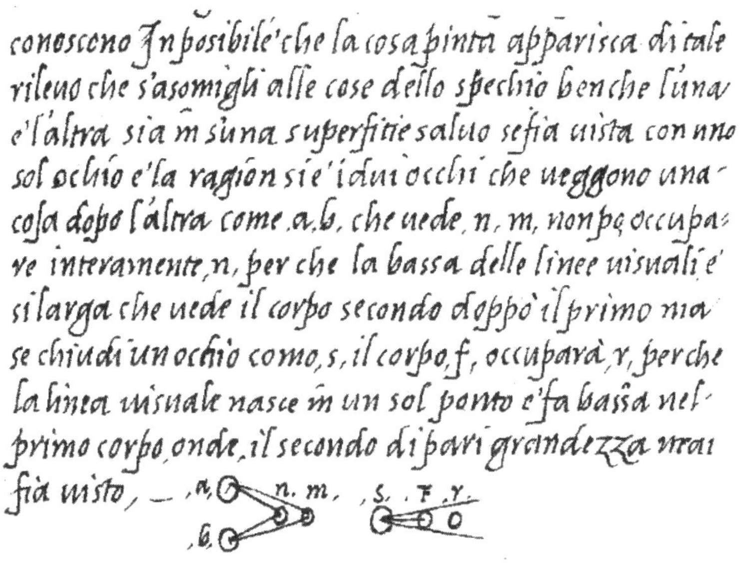

conoscono In posibile'che la cosa pinta apparisca di tale
rileuo che s'asomigli alle cose dello specchio benche l'una
e l'altra sia in suna superfitie saluo sefia uista con nno
sol ochio e la ragion sie'i dui occhi che ueggono una
cosa dopo l'altra come .a.b. che uede .n. m. non po occupa
re interamente n, per che la bassa delle linee uisuali e'
si larga che uede il corpo secondo doppo'il primo ma
se chiudi un ochio como, s, il corpo f, occupara r, perche
la linea uisuale nasce in un sol ponto e'fa bassa nel
primo corpo, onde, il secondo di pari grandezza urai
fia uisto, ___

FIGURE 1.5: From Parte Secuda (page 47) of Leonardo da Vinci's *Codex Urbinas Latinus 1270* [36].
Courtesy of Princeton University Press

(E) 1.4. This question asks about the image of circles. Figure 1.6 shows a viewer looking at three same-sized circles. Perhaps this is a top view of an artist looking at a row of columns, for example.

(a) Which of these circles has the largest *apparent size* to the viewer? (We measure apparent size by the angle that the image subtends).

(b) Which of these circles has the largest image on the picture plane?

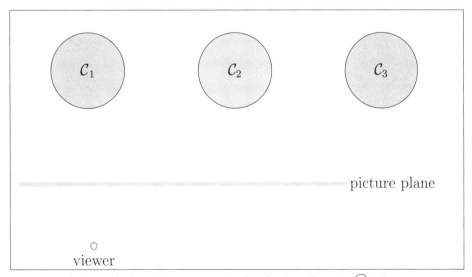

FIGURE 1.6: A viewer looks at three same-sized circles (see Exercise (E) 1.4).

(E) 1.5. Figure 1.7 shows the image of a cat on stairs, first featured on the blog 9GAG, that went "viral" in 2015 [2]. Is the cat going downstairs or upstairs? That was the question that had people clicking and sharing.

Draw two side views for this photograph:

(a) In one, assume the cat is going up the stairs,

(b) In the other, assume the cat is going down the stairs.

In each side view, keep the risers of the actual stairs vertical and the treads of the actual stairs horizontal. The main question for you to consider is that of the location and correct orientation of the picture plane.

Is this cat going

FIGURE 1.7: An internet sensation [2].

(E) 1.6. A pair of lines in $\mathbb{R}^3$ are *skew* if they neither intersect nor are parallel.

(a) Explain why two skew lines cannot lie in the same plane, using your answers to questions 1–8 from the beginning of Chapter 1.

(b) Give an example of two skew lines whose images are parallel on the picture plane.

(c) Give an example of two skew lines whose images intersect on the picture plane.

For (b) and (c), give a rough sketch of the perspective view of each scenario, drawing the picture plane, the viewer's eye, and the two planes containing the two lines respectively.

ART ASSIGNMENT

(A) 1.1. Use a digital camera or a cell phone to photograph a hallway in one-point perspective (see the Appendix below for a description of "one-point perspective"). Whether you are standing up, sitting, or lying on your belly, you should point your camera parallel to the direction of the hallway. Please do not crop the photo.

Turn in two copies of this photograph: one plain, and one that you draw on. On the second version, you will trace over several kinds of lines in the picture. (You might do this digitally, or you can use a ruler and colored pencils or colored pens.) The images of horizontal lines that run away from you (that go to a vanishing point) should be one color; the images of horizontal lines that go across the picture should be another color; and the images of vertical lines should be a third color. You should color at least five lines in each of those directions.

Extend those lines and verify for yourself that there are two sets of parallel lines and one set of lines that intersect at a point.

⚠ 1.2. Sketch your hallway in one-point perspective. Include at least two doorways, the floor and ceiling, and at least one "thing" on a wall. Make sure that lines that are supposed to be vertical in your sketch are vertical (similarly for lines that should be horizontal), and that lines that should go to the vanishing point do. Pay special attention to small details (e.g. directions of doorjambs and heights of doorknobs). Draw all lines neatly with a ruler. This sketch should take one to two hours.

PROOF/COUNTEREXAMPLE

For your write-up of your proof(s) or counterexample(s) to the statements below, you should include and refer to a side view and/or a top view.

Recall that a *jamb* is "an upright piece or surface forming the side of an opening (as for a door, window, or fireplace)" [32]; for the purposes of the statements below, we will assume that the jamb is one-dimensional, with no thickness, so that the height of the image is well defined.

You may use standard definitions and results from Euclidean geometry (for example, theorems about similar triangles).

P 1.1. Suppose that an artist sets up a vertical canvas across a sidewalk, looking down the sidewalk as it extends into the distance. The artist includes in the drawing the image of a building on one edge of this sidewalk, such that the front face of the building is parallel to the sidewalk and perpendicular to the vertical picture plane. The front of the building contains many rectangular windows that are all the same size and shape.

[T/F] Then a window jamb on the ground floor that is further down the road (and is therefore further from the artist) will have a shorter image on the picture plane than that of a jamb on a near window on the ground floor.

P 1.2. Let us assume the same conditions as in the previous statement.

[T/F] Then a window jamb on the second floor that is directly above a window on the ground floor (and is therefore further from the artist) will have a shorter image on the picture plane than that of a jamb on the ground-floor window.

Appendix
A Working Definition of *n*-Point Perspective

When we're working with the perspective image of a simple, rectangular 3-D object, the definition of the "*n*" in "*n*-point perspective" is relatively straightforward: *n* is the number of directions of edges of that object that are not parallel to the picture plane. In other words, it's the number of vanishing points of the edges.

Figures 1.8, 1.9, and 1.10 show boxes in one-point, two-point, and three-point perspective, respectively. Note that in each case, we count the vanishing points, and account for any remaining edge directions.

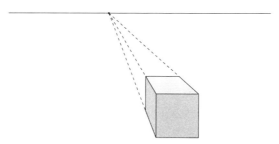

FIGURE 1.8: A box in one-point perspective. Three visible edges extend to a vanishing point, three edges run in the up/down direction, and the remaining three edges (from our viewpoint) run in the left/right direction.

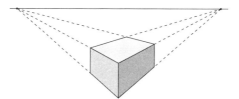

FIGURE 1.9: A box in two-point perspective. Three visible edges extend to a vanishing point, three other edges extend to another vanishing point, and the remaining three edges run in the up/down direction.

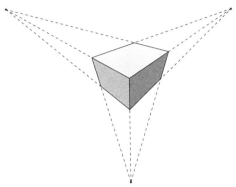

FIGURE 1.10: A box in three-point perspective. Three distinct triples of edges extend to three distinct vanishing points, accounting for all three directions.

But even within a simple box, lines that are not the edges might have other vanishing points. In fact, those other vanishing points will often be helpful in constructing our images. In general, we don't count those other vanishing points when we're describing n-point perspective. We would say that the striped box in Figure 1.11 is in "one-point perspective," even though the stripes extend to a second vanishing point, because we describe the perspective structure using the *perpendicular edges* of the box, not other lines within it.

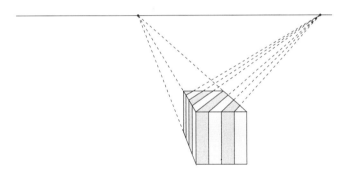

FIGURE 1.11: A box in one-point perspective. The vanishing point of the stripes doesn't contribute to the count of n for describing n-point perspective; only the edges do.

Of course, not every object in the world is a rectangular box, and so the definition of n-point perspective becomes more fuzzy and less well defined for more complicated images. A picture with several boxes might have some of the boxes in one-point perspective, some in two-point, and others in three-point perspective. Or it might have a complicated object that is not a box (a robot with arms, body, and legs all oriented in different directions). In that case, we can't say the entire picture is in n-point perspective for some n, but we might be able to say that about some parts of the picture (the torso of the robot, or one of the boxes).

Consider for example Figure 1.12. Even though the edges of the roof and the bar on the ground give us different vanishing points, because the tiled floor and the base of the house give us perpendicular lines that result in two vanishing points, artists are likely to describe the entire picture as being in two-point perspective. This is a good working description, but it's not a perfect description because the house is not a box.

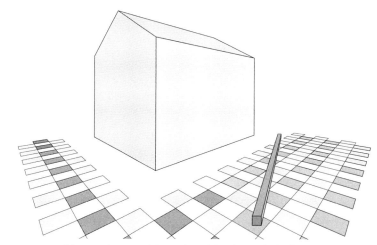

FIGURE 1.12: Most artists would say that this picture is in "two-point perspective," even though the roof and the bar on the ground give us additional vanishing points. We focus here on the tiled ground and the main part of the house.

FIGURE 2.0: The beginning of a drawing of the word ART in one-point perspective. [For use with the DRAWING ART module.]

2
Drawing ART

Overview In a previous module, we learned that some collections of parallel lines have images that are parallel on the canvas, and other collections of parallel lines have images that pass through a common vanishing point. In this module, we will use what we learned to solve some practical drawing puzzles.

In this module, we will make an 'A' in class. But first we will figure out how to make a T!

Consider Figure 2.1, which shows two attempts to complete a one-point perspective drawing of a three-dimensional T. That is, the front face of the 3-D T is parallel to the picture plane; lines going into the distance have a vanishing point on the horizon. Both drawings obviously have something wrong with them, even though the back line is vertical and the bottom line goes to the vanishing point.

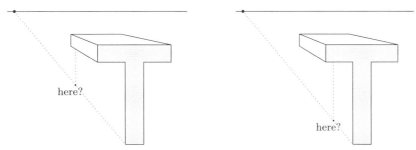

FIGURE 2.1: The base of the leftmost T is probably too long compared to the top; the base of the rightmost T is probably too short.

Your job will be to figure out construction techniques (maybe one, maybe several) for accurately finishing that picture, given a correct start to the picture. See Figure 2.2 on the next page.

1. Drawing directly on Figure 2.2, determine a method for finishing the drawing of the T.

   (A variety of correct construction techniques are possible, but few of them are immediately obvious. So feel free to take your time, doodle even. Use your straightedge, your pencil, and definitely your eraser!)

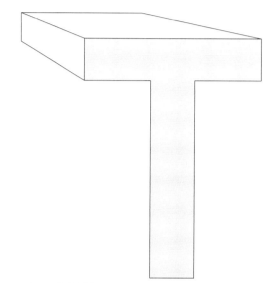

FIGURE 2.2: Draw the bottom bar of this T.

2. You might discover that your classmates have come up with a variety of clever solutions. On Figure 2.3, sketch several of these for future reference.
3. Is there one solution that strikes you as particularly "elegant"? If so, what makes that solution appeal to you?
4. What are the advantages/disadvantages of the various solutions?

Some sets of lines form natural groupings. For historical reasons, we describe the perspective image of a collection of parallel lines as a *pencil*. That is, a *pencil* of lines could be a collection of lines passing through a common point, or it could be a collection of parallel lines.

Definition A *pencil* of lines in $\mathbb{R}^3$ is a collection $\mathcal{P}$ of lines satisfying either, (a) there exists a point $P \in \mathbb{R}^3$ such that $\mathcal{P} = \{p \subset \mathbb{R}^3 : P \in p\}$, or (b) there exists a line $\ell \subset \mathbb{R}^3$ such that $\mathcal{P} = \{p \subset \mathbb{R}^3 : p \parallel \ell\}$.

Why the name "pencil"? In the 1600s, these collections of lines seemed to geometers to resemble the bristles of the brushes that they used for writing—there were no ballpoint pens or #2 pencils back then. The word for a writing brush came from the Latin for "little tail": that is, *pencil*. Only in more recent times—say, the last few hundred years—has *pencil* come to mean a graphite-based writing instrument. So when we talk about a "pencil of lines," you can think "paintbrush of lines."

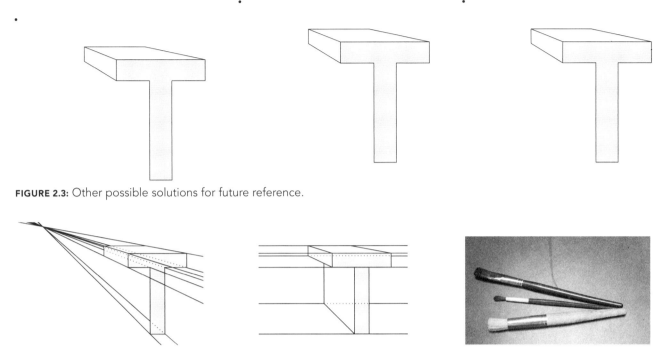

FIGURE 2.3: Other possible solutions for future reference.

FIGURE 2.4: A pencil of lines passing through a common point (left), a pencil of parallel lines (middle), and a fan-shaped paintbrush.

5. How many different pencils of lines did we use in drawing the letter T?

6. How many different lines pass through the front, top, left corner of the T?

7. How many different lines pass through the back, right, top corner of the T?

8. Explain how thinking in terms of pencils of lines can help us construct the entire T.

9. When you and the class have conquered the letter T, use your newfound mastery to complete the unfinished ART in Figure 2.0. Here are some artistic hints that you might find helpful.
   - Draw construction lines lightly.
   - Erase parts of lines you do not need.

- Draw lines with different weights: lines on the front of a letter will probably be dark; "hidden" lines could be light or dotted.
- Use a blank piece of paper on top of parts where you are not drawing (underneath your hand), so your hand doesn't smudge existing lines.
- When deciding whether something you've drawn looks "right," don't leave the paper on the desk in front of you; that's the wrong angle. Hold the paper up so your eye is level with the vanishing point; hold it far from your face so you can see the whole image all at once.

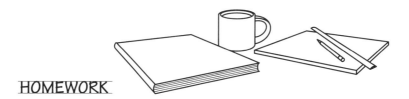

HOMEWORK

EXERCISES

(E) 2.1. How many pencils of lines are in the completed drawing of the word ART?

ART ASSIGNMENT

(A) 2.1. Draw a word that is at least four letters long in one-point perspective. The letters should be a constant width, the spaces between them should be a (smaller) constant width, and the depth of the letters should be constant (see Figure 2.5). The width of the lines in the letters should appear to be constant (the bar in a T is the same width as the bar in an "H," for example). Give the word a surrounding context that adds to the sense of depth (is it sitting on a table? mounted on the back wall of a room? in a vast plane with buildings on the horizon?)

    Draw lots of lines; draw neatly; erase no-longer-needed construction lines carefully and completely. You may be tempted to embellish your figure with decorations. You should do so if these provide additional perspective context—in fact, the more perspectively correct embellishments there are, the better the picture will look. But you must use a straightedge for *every* line you draw, and you should not draw any curves.

(A) 2.2. Find a rectangular solid in your environment (for example, a tissue box or a bookshelf). Take three photographs of this object, each showing three faces of the object. In particular, each of the three photographs should show the images of nine edges of the object (that is, the images of three vertical edges, three horizontal edges heading one direction, and three horizontal edges heading another direction).

- One photograph should be in one-point perspective, meaning one face of the object should be parallel to the picture plane and one set of edges of your object will appear to converge to a vanishing point.[1]

---

[1] Hint: The vanishing point should be straight ahead, so if your photograph is uncropped, the vanishing point will be exactly in the middle of the picture.

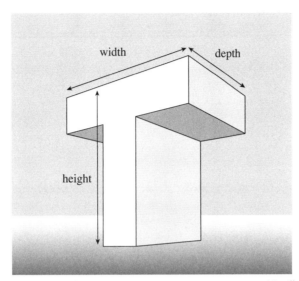

FIGURE 2.5: This T is not in one-point perspective. You'll learn about two-point perspective in future assignments.

- A second photograph should be in two-point perspective, meaning one edge of the object should be parallel to the picture plane and two sets of edges will appear to converge to vanishing points.
- The third photograph should be in three-point perspective, meaning no edge of the object should be parallel to the picture plane.

Either digitally, or else by printing and drawing on the copies with rulers and colored pencils, you should extend the nine visible edges of your object and determine the location of the vanishing point(s), or else verify that the edges have parallel images. For the two- and three-point perspective pictures, you may need to shrink the picture or attach extra paper, because the vanishing points could possibly be off the edges of the picture.

PROOF/COUNTEREXAMPLE

[P] 2.1. In this problem, we consider distinct pencils $\mathcal{L}$ and $\mathcal{K}$ in $\mathbb{R}^3$. ("Distinct" means that these are not the same pencils, that is, that there is at least one line in $\mathbb{R}^3$ that is in one pencil but not in the other).

Give a true statement about $\mathcal{L} \cap \mathcal{K}$, the set of lines contained in both pencils, and then prove your statement. Your statement and proof might include two or more cases.

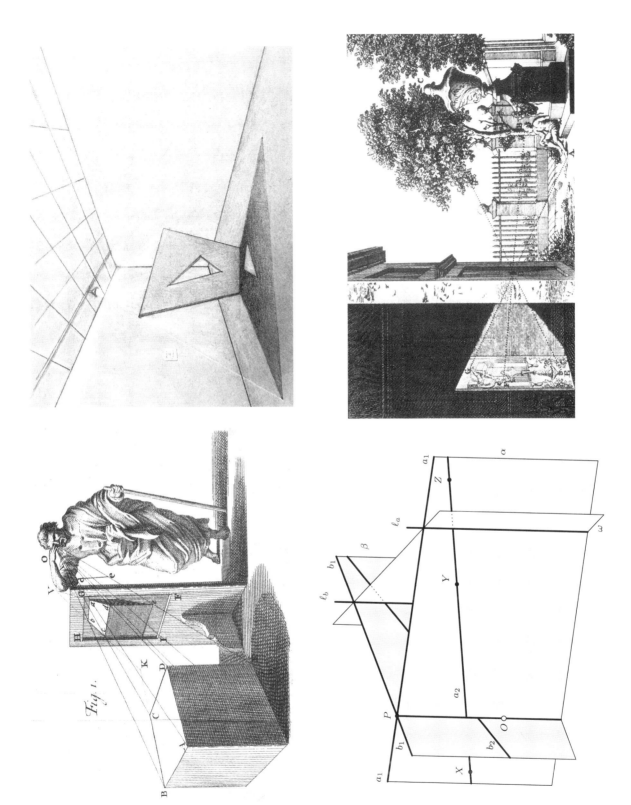

FIGURE 3.0: Linear perspective setup from [51]; study for the cover of our book by Fumiko Futamura; three vertical planes α, β, and ω; a camera obscura as illustrated in the *Universal Magazine* [1]. [For use with the WHAT'S THE IMAGE OF A LINE? module.] (Brook Taylor piece): Courtesy of the Max Planck Institute for the History of Science, Berlin (Camera Obscura): Wellcome Collection

3

What's the Image of a Line?

Overview When we taped windows, we had an artist, and then a picture plane, and then the object, in that order. But that configuration is not the only possible way to project the image of an object onto a plane. Figure 3.0 shows that there are many possible kinds of projections. In this module, we'll explore many possible notions of "projection onto a plane." Our goal will be to try to create a more general mathematical model for projection that might include any or all of these types of projections.

Let us think about possible ways of projecting the images of points and lines onto a plane. Every projection we consider contains

- an *object* being projected,
- an *image* of the projection, and
- a point that we will call the *center* from which the object and image line up with each other.

1. The top left drawing of an artist drawing a cube in Figure 3.0 is from Brook Taylor's *New Principles of Linear Perspective* [51], published in 1719. (Some math students will have seen a calculus concept called "Taylor series"; this is the same Taylor). In this drawing, we see a projection like the one we did ourselves in class. In this drawing, what is the *object*? What is the *image*? What is the *center* of the projection?
2. The top right drawing, inspired by the art of Roger Jorgensen, is a study by Futamura for the cover of our book. It shows, among other things, the shadow of a Jorgensenesque geometric sculpture. If we think of a shadow as a projection, then what is the *object* of the projection? What is the *image*? What is the *center* of the projection?
3. The lower right drawing in Figure 3.0 shows a *camera obscura*—translated literally as "dark room." In this drawing, the projection is caused by light rays passing from outside of the room and being cast upon a wall inside the room. What is the *object*? What is the *image*? What is the *center* of the projection?

In this class (as elsewhere in Projective Geometry), we will use the convention of naming points with italicized capital letters, lines with italicized lowercase letters, and planes with lowercase Greek letters. Unless we state otherwise, when we project objects onto a plane we will let O denote the *center of the projection* and ω denote the picture plane. For any object X, we will let X' be the image of X in the picture plane ω.

We should be cautious here. We have not yet defined exactly what we mean by "the projection via O onto the picture plane ω," and (as you will see below) there are several different things we might mean by this phrase. For now we'll leave the meaning ambiguous.

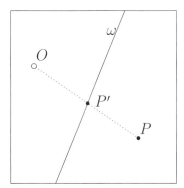 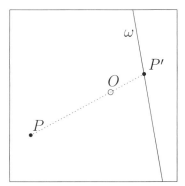

FIGURE 3.1: Two versions of a projection: on the left, as with an artist looking though a window; on the right, as with a camera projecting onto film. In each version, O is the center of the projection, ω is the picture plane, and the point P projects to its image P'.

4. In Figure 3.1, we see that the image of a point could be another point. That is, if P is a point, P' could be a point.
   (a) Could we have $P' = \varnothing$? (The symbol $\varnothing$ is the *empty set*, the set that contains nothing.)

   (b) Could P' be a line?

   (c) Could P' be a plane?

   For each of your answers above, draw a top view and side view and/or a 3-D sketch explaining your reasoning.

5. We know that the image of a line in $\mathbb{R}^3$ could be another line. The physical world does not always correspond to the abstract mathematical setting, however. Draw top and side views and/or 3-D sketches for a geometrical setting, or give physical examples from a real-world setting, to show whether the image ℓ' could take the following forms:

(a) $\varnothing$;

(b) a point;

(c) a line segment;

(d) a ray;

(e) a line with one point missing;

(f) a line with two points missing; or

(g) an ellipse.

Eventually, for reasons that will make our mathematical lives easier, we will want to be able to assume that the image of a point is always a point and that the image of a line is always a line. That is, eventually we will need a way to rule out any of the counterexamples you might have discovered above. But that will be the topic of a future module.

Meanwhile, here are some questions that might help you visualize projections in three-dimensional space.

6. Consider Figure 3.2, which shows two intersecting vertical planes α and β, which meet a vertical picture plane ω in parallel lines ℓ_a and ℓ_b, respectively. From the center O, we project objects in $\mathbb{R}^3$ to their images in ω.

(a) The line a_2 (containing the points X, Y, and Z) lies in plane α. Locate the perspective images X', Y', and Z' of points X, Y, and Z, respectively. (*Hint*: To locate X' you may have to extend a line of the drawing.)

(b) What is the perspective image of the entire line a_2?

(c) The line b_2 lies in plane β. What is the image of the entire line b_2?

(d) Are a_2 and b_2 parallel? Do they intersect? How can you tell?

(e) What is the image of a_1? What is the image of b_1?

(f) The two lines a_1 and b_1 meet at point P. What is the relationship of line OP to the picture plane ω?

(g) If a line lies in plane α and does not pass through O, its perspective image is _____. If a line lies in plane β and does not pass through O, its perspective image is _____.

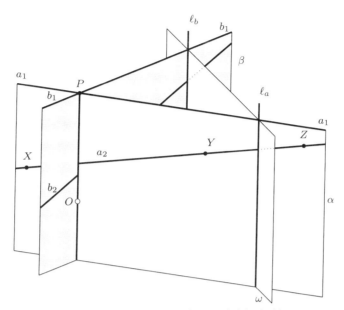

FIGURE 3.2: Three planes (ω, α, and β), with labeled lines (a_1, a_2, b_1, b_2, ℓ_a, and ℓ_b) and points (O, P, X, Y, and Z), for question 6.

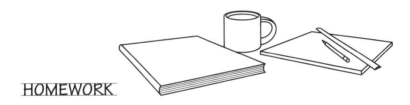

HOMEWORK

(E) 3.1. Figure 3.3 shows an aerial view of a fenced-in area. Around this area there are low stone walls, and at each of the four corners is a flag on a tall post. Each flag has one of four letters (P, O, S, or T).

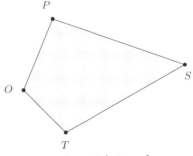

FIGURE 3.3: An aerial view of a fenced-in area.

Figure 3.4 shows how a person standing at point X outside the stone walls would read the flags, from left to right, as "STOP," whereas a person standing at point Y would read the flags from left to right as "PSOT." What other "words"

4

The Geometry of $\mathbb{R}^2$ and $\mathbb{R}^3$

4.1 Euclidean Geometry: A Point of Comparison

Overview We've already been explicitly and implicitly using concepts from Euclidean geometry—concepts such as parallel lines, similar figures, and congruent figures. This module will help us solidify some of these concepts before we dig deeper into projective geometry.

1. Try to write down your own definition of a "point." Then look at the key words you used to define a point, and define those words. Can you define "point" without using the notion of point?

2. Can you define "line" without using the notion of line?

You may have had difficulty avoiding defining the words above with the words themselves. "Points" and "lines" are notions we understand intuitively that have no definition based on other, previously defined, mathematical objects. We call these *primitive notions*.[1] A "plane" is another primitive notion. Not only objects, but also relationships between objects, can be primitive notions; for example, we will see in the questions below the concepts of *betweenness*, *incidence*, and *congruence*.

3. Give some examples of what we might mean by *betweenness*, and illustrate the examples with pictures.

---

[1]We will see later with the Principle of Duality in Projective Geometry that it really does help to think of the primitive notions of point and line as undefined, and not even attached to a visual image.

4. *Incidence* refers to meeting or touching. A point is *incident* with a line if the point is on the line. A line is *incident* with a point if it goes through the point. Give some other relationships involving point, line, and plane using the word "incident," and illustrate the relationships with pictures.

5. *Congruence* intuitively means sameness of measurement, but not necessarily in the same place. Explain what we might mean when we say two angles are *congruent*, or for a geometric object to be congruent to another geometric object.

From our primitive notions of the most basic objects and relationships, we can develop rules and more complicated definitions. A rule, also called an *axiom* or *postulate*, is a statement we assume to be true and do not therefore prove. One basic axiom of Euclidean space is the following:

Axiom 1 (Two points determine a line). *Two distinct points determine a unique line. In other words, for any two distinct points A and B, there is one and only one line incident with both A and B.*

6. Is it also true that in Euclidean geometry, "two lines determine a point"? In other words, does every pair of distinct lines meet at a point?

Saying "two lines don't always determine a point" is too vague for formal mathematics, so we rephrase the idea in the form of the following axiom.

Axiom 2 (Parallel Postulate). *Given a line and a point in a plane, the point not incident with the line, there exists a unique line in the plane that is incident with the point but not incident with the line. That is, there is a unique line in the plane that is parallel to the first line and that passes through the point.*

Euclid's original Parallel Postulate made less intuitive sense than Axiom 2, often called "Playfair's Axiom" after the mathematician John Playfair. Axiom 2 better clarifies what we mean when we say that there are lines that are not incident. But the axiom would be even easier to understand if we gave the objects specific names.

For this reason, we will use the following convention for naming objects. We will name points with capital Roman letters (A, B, etc.); we will name lines with lowercase Roman letters (ℓ, k, etc.) or, using the first axiom, as pairs of points ($\ell = AB$); we will name planes with lowercase Greek letters (ω, ϕ, etc.). We will denote the intersection of two lines by a dot ($\ell \cdot m$).

In the remainder of this module, we assume everything we discuss occurs in a plane, so that we are focusing on two-dimensional Euclidean geometry. Notice how the following definitions use the primitive notions of line, point, incidence, and betweenness.

Definition If two or more points are incident with a common line, we say they are *collinear points.*

If two or more lines are incident with a common point, we say they are *concurrent lines.*

A *line segment* is the section of a line that is between two distinct points on the line, called *endpoints.* We denote a line segment as we would a line, except with a line segment over it; by $\overline{AB}$.

A *ray* is the section of the line that has only one endpoint and extends infinitely in one direction, denoted by $\overrightarrow{AB}$, where A is the endpoint and B is any point along the ray.

Two rays $\overrightarrow{AB}$ and $\overrightarrow{AC}$ (or line segments $\overline{AB}$ and $\overline{AC}$) that are incident at endpoint A form two sections between the rays called *angles with vertex A*, both unfortunately denoted by $\angle BAC$. (When we say "sections," we mean that if we start at ray $\overrightarrow{AB}$, we can rotate the ray about the point A in two different directions, clockwise and counterclockwise to reach ray $\overrightarrow{AC}$. These turns sweep out two different sections of the plane. Usually, smaller of the turns is what we mean by $\angle BAC$.)

7. Draw properly labeled examples of the following:
   (a) collinear points
   (b) concurrent lines
   (c) a line segment
   (d) a ray
   (e) an angle

Definition A *triangle* is a geometric figure made of three non-collinear points and the three lines determined by pairs of the points.

A *quadrangle* is a geometric figure made of four points, no three of which are collinear, and the six lines determined by pairs of the points.

Why does the definition of a quadrangle say it has "six" lines instead of "four" lines? After all, when we think of a four-sided shape, we usually don't consider the diagonals. The reason for including all six lines is partly because of the parallel language we used in the definition of a triangle and a quadrangle. Moreover, as we have seen in drawing in perspective, diagonals are useful and important.

that a pair of angles whose measurements sum to 180 degrees are *supplementary angles*. For example, in Figure 4.1, ∠3 and ∠ 4 are supplementary.

A transversal of two lines is a line that crosses both lines. In Figure 4.1 we demonstrate other relationships between pairs of angles in this context:

- ∠1 and ∠3 are *corresponding angles*, as are ∠4 and ∠5,
- ∠1 and ∠2 are *vertical angles*,
- ∠2 and ∠3 are *alternate interior angles*, and
- ∠2 and ∠4 are *interior angles*.

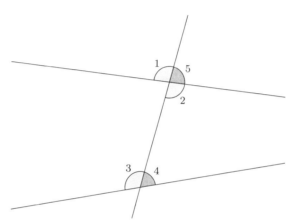

FIGURE 4.1: Transversal

We mentioned earlier that Euclid's original Parallel Postulate was not very intuitive. But it might come in useful here.

Euclid's Parallel Postulate: If a line segment intersects two straight lines forming two interior angles on the same side that sum to less than two right angles, then the two lines, if extended indefinitely, meet on the side on which the angles sum to less than two right angles.

12. If the two lines, when extended indefinitely, do *not* meet on one side of a transversal, then what can we conclude about the sum of the two interior angles on that side?

13. If two lines, when extended indefinitely, do not meet on *either* side of a transversal, then what can we conclude about the sum of the two interior angles on one side of the transversal?

14. Prove that if a line is cut by a transversal, then the vertical angles are congruent.

15. Prove that if a pair of parallel lines is cut by a transversal, then the alternate interior angles are congruent.

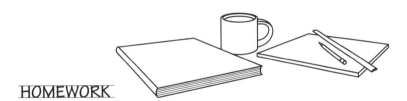

HOMEWORK

[P] 4.1.1. Prove that if A, B, C, and D are collinear in that order and $\overline{AB} \cong \overline{CD}$, then $\overline{AC} \cong \overline{BD}$.

[P] 4.1.2. Prove that if a pair of parallel lines is cut by a transversal, the corresponding angles are congruent.

[P] 4.1.3. Prove that the measurements of the three interior angles of a triangle sum to 180 degrees.

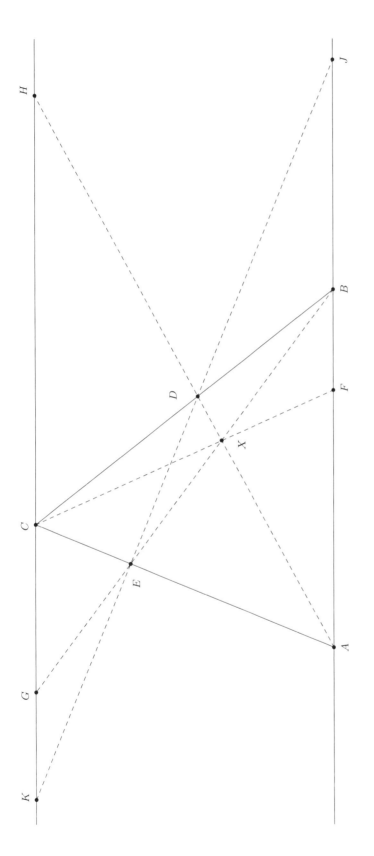

FIGURE 4.2.0: The triangle $\triangle ABC$ with a few other lines added. [For use with the EUCLIDEAN GEOMETRY REVISITED: SIMILARITIES AND INVARIANTS module.]

2. (SAA) If two corresponding angles and a side not between them are congruent, then the triangles are congruent.

3. (SSA) If two corresponding sides and an angle not between them are congruent, then the triangles are congruent.

4. (AAA) If the three corresponding angles of two triangles are congruent, then the triangles are congruent.

5. If we define an *isosceles triangle* to be a triangle with two congruent sides, prove that the angles opposite these congruent sides are congruent.

You may have found that AAA does not guarantee congruence. Even though two triangles with the same angles as each other might not be congruent, they do still have a special relationship, in that one is merely a smaller or larger version of the other. We call them *similar triangles*.

Axiom 9 (Similarity). *Two triangles $\triangle ABC$ and $\triangle A'B'C'$ satisfy (AAA) if and only if their corresponding sides are proportional by some linear scaling factor; that is, there exists $k > 0$ such that $\|AB\| = k \cdot \|A'B'\|$, $\|BC\| = k \cdot \|B'C'\|$, and $\|AC\| = k \cdot \|A'C'\|$.*

6. If one triangle has sides of length 3, 4, and 5, and the other has sides of length 6, 8, and 10, are these triangles similar? If so, what is k?
7. If one triangle has sides of length 3, 3, and 5 and the other has sides of length 6, 10, and 10, are these triangles similar? If so, what is k?

Ceva's and Menelaus's theorems

Using the ideas of similarity and parallelism, we are able to prove two very interesting theorems involving directed distances around a triangle.[2] We have already talked about distance (length). What is a directed distance?

A **directed distance** is exactly what it sounds like: a distance with direction. Imagine you are standing at 0 on the number line. If you go in the positive direction, then the directed distance between where you were and where you end up is positive. If instead, you go in the negative direction, then the directed distance is negative. Directed distance is helpful when you want to compute *displacement* as opposed to *total distance traveled*.

8. If you start at 0 and go in the positive direction 5 units, and from there you go in the negative direction 3 units, then
    (a) your total distance traveled is _____
    (b) and your displacement from 0 is _____
9. By adding directed distances, we obtain _____

[2]These are pronounced "CHAY-vuh" and "Meh-nuh-LAY-us."

We denote directed distance by $|AB|$. Thus, if A is at $x = 2$ and B is at $x = 7$, then $|AB| = 5$ and $|BA| = -5$.

Using directed distances, we are able to prove two famous results about triangles. No matter what the size or shape of the triangle, there are two interesting constants having to do with looking at ratios of divided side lengths. In the steps that follow, you will discover the value of these interesting constants.

Ceva's Theorem. *Let $\triangle ABC$ be a triangle, and let D, E, and F be on the lines BC, CA, and AB respectively such that lines AD, BE, and FC are concurrent. Then*

$$\frac{|AF|}{|FB|} \cdot \frac{|BD|}{|DC|} \cdot \frac{|CE|}{|EA|} = \text{———}.$$

10. Figure 4.3 shows a setup for Ceva's theorem that has the point of concurrence inside the triangle. Draw an alternative setup for Ceva's theorem, this one showing the point of concurrence *outside* the triangle.

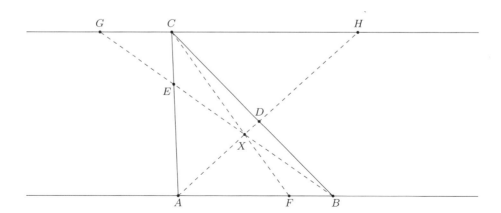

FIGURE 4.3: Shade the appropriate triangles, for question 11.

Let's try to think about how we might discover the missing quantity in Ceva's theorem. Ceva's theorem requires us to find a relationship among the ratios of sections of sides of triangle $\triangle ABC$. We know relationships between ratios of sides that involve similar triangles, which requires us to find two triangles with corresponding angles congruent. There do not seem to be any similar triangles in the original setup of Ceva's theorem; we will create similar triangles ourselves.

In the first Euclidean Geometry module, we showed that if a pair of parallel lines is cut by a transversal, then we can identify congruent angles such as pairs of vertical angles, alternate interior angles, and corresponding angles. So we draw a line through C parallel to side AB, as shown in Figure 4.3, then let G and H denote the intersection of that line with the lines BE and AD respectively.

11. In Figure 4.3, shade in a pair of similar triangles such that one of the triangles has a side AF. Be careful how you order the points in the triangle as you name them.
12. Explain why these triangles are similar, using our axioms and previously proven theorems.

13. Because these triangles are similar, we have three equal ratios:

$$\frac{|AF|}{(\quad)} = \frac{(\quad)}{(\quad)} = \frac{(\quad)}{(\quad)}.$$

14. In Figure 4.3, shade in a pair of similar triangles such that one of the triangles has a side FB (make the shading a little different from the shading for question 11). Be careful how you order the points in the triangle as you name them.
15. Is the reason these triangles are similar the same as that for question 12?
16. Because these triangles are similar, we have three equal ratios:

$$\frac{|FB|}{(\quad)} = \frac{(\quad)}{(\quad)} = \frac{(\quad)}{(\quad)}.$$

17. What sides do the two pairs of similar triangles have in common?
18. Use your answers to questions 13 and 16 to determine a simple ratio equivalent to $|AF|/|FB|$:

$$\frac{|AF|}{|FB|} = \frac{(\quad)}{(\quad)}.$$

You should now have a relationship between the lengths $|AF|$ and $|FB|$ and two line segments that are not on the original triangle. Our next task is to get rid of these non-triangle quantities by replacing them with ratios involving the remaining four line segments.

19. In Figure 4.4, shade in a pair of similar triangles such that one triangle has a side BD and the other triangle has a side CD. Be careful how you order the points in the triangle as you name them.
20. Is the reason these triangles are similar the same as that for question 12?

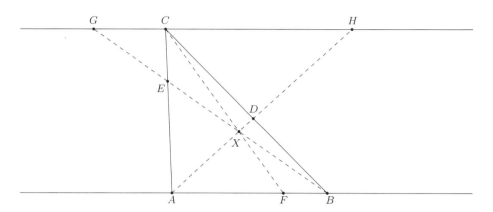

FIGURE 4.4: Shade the appropriate triangles, for question 19.

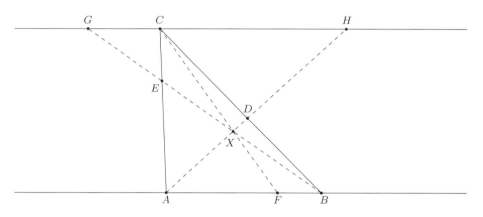

FIGURE 4.5: Shade the appropriate triangles, for question 22.

21. Because these triangles are similar, we have three equal ratios:

$$\frac{|CD|}{(\quad)} = \frac{(\quad)}{(\quad)} = \frac{(\quad)}{(\quad)}.$$

22. In Figure 4.5, shade in a pair of similar triangles such that one triangle has a side *CE* and the other triangle has a side *AE*. Be careful how you order the points in the triangle as you name them.

23. Is the reason these triangles are similar the same as that for question 12?

24. Because these triangles are similar, we have three equal ratios:

$$\frac{|AE|}{(\quad)} = \frac{(\quad)}{(\quad)} = \frac{(\quad)}{(\quad)}.$$

25. Combine the equations from questions 18, 21, and 24 to get the conclusion of Ceva's theorem. You may find that some of the letters are flipped (perhaps in one place, you have |CD| and in another you have |DC|). Remember that when you flip

the order, because they are directed distances, you should include a negative sign: $|CD| = -|DC|$.

The converse of Ceva's theorem turns out to also be true. You will prove this in the homework.

Converse of Ceva's Theorem. *Let $\triangle ABC$ be a triangle, and let D, E, and F be on the lines BC, CA, and AB respectively. If*

$$\frac{|AF|}{|FB|} \cdot \frac{|BD|}{|DC|} \cdot \frac{|CE|}{|EA|} = \underline{\qquad},$$

then the lines AD, BE, and FC are concurrent.

A related theorem looks at a line *JK* that intersects each of the three sides at three distinct points, none of which are the vertices.

Menelaus's Theorem. *Let $\triangle ABC$ be a triangle, and let a transversal line intersect sides BC, CA, and AB at D, E, and J respectively such that D, E, and J are distinct from A, B, and C. Then*

$$\frac{|AJ|}{|JB|} \cdot \frac{|BD|}{|DC|} \cdot \frac{|CE|}{|EA|} = \underline{\qquad}.$$

26. Figure 4.6 shows a setup for Menelaus's theorem that has the point *D* between *B* and *C*. Draw an alternative setup for Menelaus's theorem, this one showing the point *D not* between *B* and *C*. (You will prove Menelaus's theorem in your homework. The proof is similar to that of Ceva's theorem.)

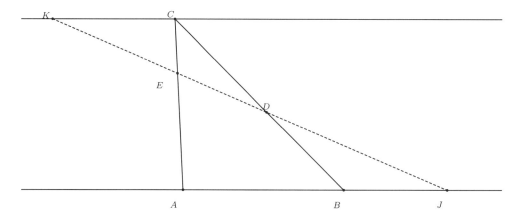

FIGURE 4.6: One possible setup for Menelaus's theorem.

HARMONIC SET

Now consider Ceva's and Menelaus's theorems superimposed on top of each other as shown in Figure 4.7. Notice, Ceva creates the point F on the line AB and Menelaus creates the point J. Using the two theorems, we are able to come up with the following relationship between these four points along a line, called the *harmonic set*.

> **Definition** Let $\triangle ABC$ be a triangle, and let a transversal line intersect sides BC, CA, and AB at D, E, and J respectively such that D, E, and J are distinct from A, B, and C. Denote the intersection points by $X = AD \cdot BE$ and $F = CX \cdot AB$. Then we say the points $AFBJ$ form a *harmonic set* (which we denote by H(AB, FJ)).

Harmonic Ratio Theorem. *Let A, F, B, and J be four points along a line, in that order. This set of points is a harmonic set if and only if*

$$\frac{|AF|}{|FB|} \cdot \frac{|BJ|}{|JA|} = -1.$$

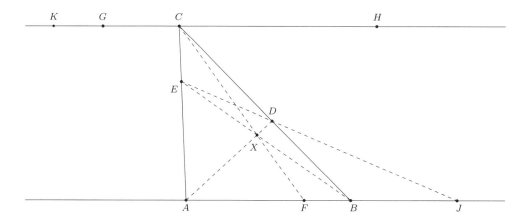

FIGURE 4.7: The harmonic set H(AB, FJ).

27. Figure 4.2.0 shows a triangle $\triangle ABC$ with both Ceva's and Menelaus's config-
    urations super-imposed. But this figure also relates to perspective lessons we will
    see again, especially when we get to the sections on drawing cubes in two-point
    perspective and on numerical projective invariants.

    To locate the hidden perspective image in the triangle, shade the quadrangle
    CEXD in Figure 4.2.0. Then turn the picture upside down and think of the
    quadrangle as representing the image of a rectangle on the ground.

    (a) If this diagram now represents a perspective image, what does the line AB
        represent?

    (b) What role do the points A and B now play?

    (c) What do the points F and J represent?

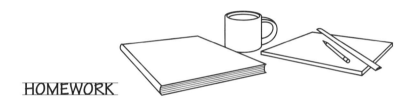

HOMEWORK

PROOF/COUNTEREXAMPLE

P 4.2.1. Prove the converse of Ceva's theorem. (*Hint*: Let $X = AD \cdot BE$ and $F' = CX \cdot AB$.
Use Ceva's theorem to say something about the situation involving F', then use
that to show $F = F'$.)

P 4.2.2. Prove Menelaus's theorem, and while doing so, determine the constant that
goes in the blank. (*Hint*: As in the proof of Ceva's theorem, draw a line through
C parallel to AB and find similar triangles again. See Figure 4.6.)

P 4.2.3. Using Ceva's and Menelaus's theorems, prove the harmonic ratio theorem.

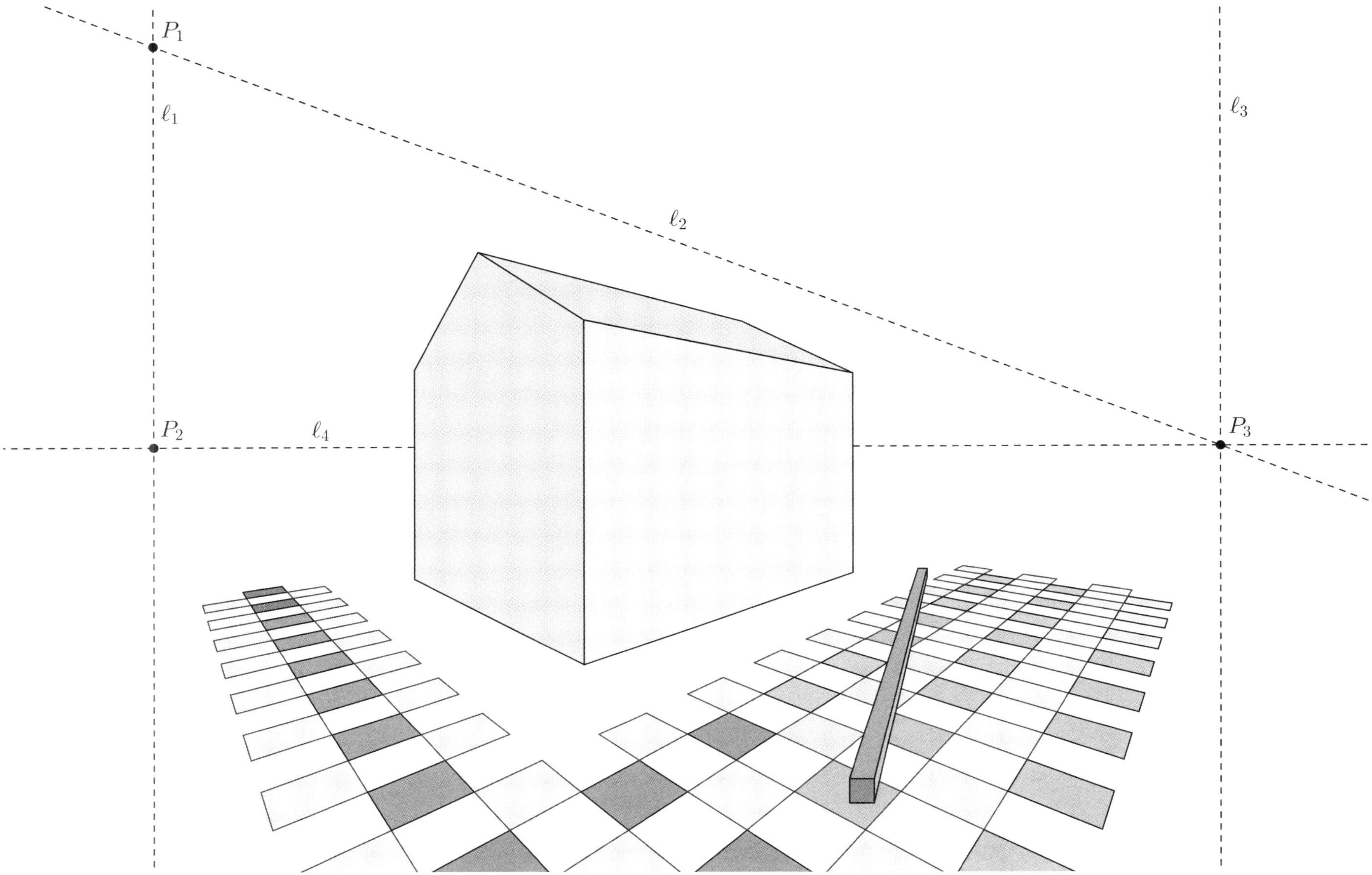

FIGURE 5.0: Extending the space we live in. [For use with the EXTENDED EUCLIDEAN SPACE: TO INFINITY AND BEYOND module.]

5

Extended Euclidean Space

To Infinity and Beyond

Overview Because of the issues we discovered in the IMAGE OF A LINE module, attaching some additional points to $\mathbb{R}^3$ will ensure we don't have to keep saying things like,

> "If these two things are parallel, then something happens, but if they're not parallel, then something else happens."

Adding these new points "extends" Euclidean space. We'll develop a set of definitions for points, lines, and planes in extended Euclidean space, and then we'll use these definitions to start doing formal proofs about what happens in this new, larger space.

How can we visualize a space that includes all the points we need—even the points that have "vanished"? If we think of $\mathbb{R}^2$ as the ground plane, then a perspective drawing of $\mathbb{R}^2$ (such as the checkered floor in Figure 5.0) often includes images of the collection of points "at infinity," that is, points on the horizon. Lines in the drawing that meet at a point on the horizon (for example, the lines that meet at P_2) are the images of parallel lines, so it makes sense to declare that, for every collection of parallel lines in $\mathbb{R}^2$, they intersect at a "point at infinity" in some larger, extended space, which we will call $\mathbb{E}^2$.

To be able to prove things mathematically, we will need precise definitions. The definitions below allow us to formally investigate the properties of $\mathbb{E}^2$; these definitions emphasize that "points at infinity" (or, more formally, "ideal points") are intimately connected with sets of lines that connect to them.

Definition For any line $\ell \subset \mathbb{R}^2$, we denote the set of all lines parallel to ℓ by the symbol $[\![\ell]\!]$. We will then define a new object, denoted by $P_{[\![\ell]\!]}$, which we call the *ideal point* of ℓ.

> (This notation is meant to remind us of a combination of "{}" and "||," the symbol for set and the symbol relating parallel lines: for example, "$\ell||k$" means ℓ is parallel to k).

Note that even though we call $P_{[\![\ell]\!]}$ a "point," it is not actually a point in $\mathbb{R}^2$. The definition above is different from the one defining, for example, a "vanishing point" of ℓ, where we declared which of the points already in existence in $\mathbb{R}^3$ lies in the plane ω on a line of sight parallel to ℓ. In contrast, an *ideal point* brings something into $\mathbb{E}^2$ that does not exist in $\mathbb{R}^2$. Sometimes mathematicians call this a "formal definition," because it *forms* a new object.

The next set of definitions tells us how ideal points "connect" to points and lines in $\mathbb{R}^2$.

Definition The *extended plane*, $\mathbb{E}^2$, consists of the points of the Euclidean plane $\mathbb{R}^2$ together with the collection of ideal points of lines in $\mathbb{R}^2$ such that the following conditions hold.

- Elements of $\mathbb{E}^2$ are *points*; a point in $\mathbb{E}^2$ is either an *ordinary* point $P \in \mathbb{R}^2$ or an *ideal* point $P_{[\![\ell]\!]}$ for some line $\ell \subset \mathbb{R}^2$.
- A line in $\mathbb{E}^2$ is either the *ideal line* ℓ_∞ (which we define to be the union of all ideal points in $\mathbb{E}^2$) or an *ordinary line* $\ell = \ell_0 \cup P_{[\![\ell_0]\!]}$ (obtained from the union of the points of a Euclidean line ℓ_0 together with the ideal point $P_{[\![\ell_0]\!]}$ of that line.)

---

1. Think of the ground plane in Figure 5.0 as $\mathbb{E}^2$. That is, ignore lines in the house for now, and pay attention only to lines on the ground. Which sets of lines in the ground plane of Figure 5.0 contain the ideal points named below:
   - the ideal point whose perspective image is P_2?
   - the ideal point whose perspective image is P_3?
2. In Figure 5.0, somebody left a long bar lying on the ground. Where is the image of the ideal point for the long edges of this bar?

What can we conclude about the space $\mathbb{E}^2$? Let's explore. When you see a statement listed as [T/F], you should decide whether the statement is true or false, and then provide a convincing argument that you are right. You should support your arguments using the definitions above and the already-proved statements from previous chapters.

3. [T/F] There exists a pair of distinct lines $k, \ell \in \mathbb{E}^2$ with $k \cdot \ell = \emptyset$.

   ["$k \cdot \ell$" is the intersection of lines k and l.]

4. [T/F] There exists a pair of distinct lines $k, \ell \in \mathbb{E}^2$ such that $k \cdot \ell$ contains a single point.

5. [T/F] It is possible for a pair of distinct lines in $\mathbb{E}^2$ to have two or more points of intersection.

6. [T/F] Every pair of distinct lines in $\mathbb{E}^2$ intersects in one and only one point.

7. [T/F] If P is an ordinary point and Q is an ideal point, then PQ exists.

["PQ" is the line containing points P and Q.]

8. [T/F] Every pair of distinct points P and Q in $\mathbb{E}^2$ determines exactly one line. (Here, "determines" means there is ____ and only ____ line that contains both ____ and ____.)

9. The points P_2 and P_3 in Figure 5.0 lie on the horizon, which is the image of the ideal line in $\mathbb{E}^2$. How do we know from the definitions above that P_2 and P_3 must each also lie on the images of ordinary lines, as well?

In the same way that we defined the extended plane as an extended Euclidean plane, we may define *extended Euclidean space*, $\mathbb{E}^3$, as an extension of $\mathbb{R}^3$, as in the definitions below.

Definition For any plane $\alpha \subset \mathbb{R}^3$, we denote the set of all planes parallel to α by the symbol $[\![\alpha]\!]$. We will then define a new object, denoted by $\ell_{[\![\alpha]\!]}$, which we call the *ideal line* of α. This ideal line contains a collection of ideal points: we say $P_{[\![k]\!]} \in \ell_{[\![\alpha]\!]}$ whenever the Euclidean line k is parallel to the plane α.

Definition *Extended space*, $\mathbb{E}^3$, consists of the points of Euclidean space $\mathbb{R}^3$ together with the collection of all ideal points of lines in $\mathbb{R}^3$ such that the following conditions hold.

- Elements of $\mathbb{E}^3$ are *points*; a point in $\mathbb{E}^3$ is either a point in $P \in \mathbb{R}^3$ (called an *ordinary point*) or an ideal point $P_{[\![\ell]\!]}$ for some line $\ell \subset \mathbb{R}^3$.
- A line in $\mathbb{E}^3$ is either an *ideal line* $\ell_{[\![\alpha]\!]}$ for some plane $\alpha \subset \mathbb{R}^3$ or an *ordinary line* $\ell = \ell_0 \cup P_{[\![\ell_0]\!]}$ (obtained from the union of the points of a Euclidean line $\ell_0 \subset \mathbb{R}^3$ together with the ideal point $P_{[\![\ell_0]\!]}$ of that line.)
- A plane in $\mathbb{E}^3$ is either the *ideal plane* α_∞ (which we define to be the union of all ideal points in $\mathbb{E}^3$) or an *ordinary plane* $\alpha = \alpha_0 \cup \ell_{[\![\alpha_0]\!]}$ (obtained from the union of the points of a Euclidean plane $\alpha_0 \subset \mathbb{R}^3$ together with the points in the ideal line $\ell_{[\![\alpha_0]\!]}$ of that plane.)

As before, we will explore both the mathematical and the artistic implications of these definitions, beginning with mathematical implications.

10. [T/F] Every pair of distinct points in $\mathbb{E}^3$ determines exactly one line.

11. [T/F] Any three points in $\mathbb{E}^3$ determine exactly one plane.

12. [T/F] Any two distinct intersecting ordinary lines determine exactly one plane.

13. [T/F] An ordinary line together with an ideal line that share a common point determine exactly one plane.

14. [T/F] Any two distinct intersecting ideal lines determine exactly one plane.

15. [T/F] Any two distinct ideal lines in $\mathbb{E}^3$ intersect in one and only one point.

16. [T/F] Any two distinct lines in $\mathbb{E}^3$ intersect in one and only one point.

17. [T/F] If a pair of distinct lines in $\mathbb{E}^3$ is coplanar, then they intersect in a point.

18. [T/F] Two distinct lines in $\mathbb{E}^3$ intersect in a point only if the lines are coplanar.

19. [T/F] If the line $\ell \subset \mathbb{E}^3$ is not a subset of the plane $\alpha \subset \mathbb{E}^3$, then $\ell \cdot \alpha$ is exactly one point.

20. [T/F] Any two distinct planes in $\mathbb{E}^3$ intersect in exactly one line.

21. [T/F] Any three planes in $\mathbb{E}^3$ intersects in exactly one point.

Questions 22–25 refer again to Figure 5.0.

22. In Figure 5.0, which sets of planes contain the ideal lines named below:
    - the ideal line whose perspective image is ℓ_1?
    - the ideal line whose perspective image is ℓ_2?
    - the ideal line whose perspective image is ℓ_3?
    - the ideal line whose perspective image is ℓ_4?

23. Locate the image of the ideal line for the back slanted face of the roof.

24. Using a straightedge and pencil, add additional detail to the image of the house in Figure 5.0 (you might use Figure 5.1 as a prototype, but you should draw directly on Figure 5.0). Include at the very least these objects:
    - a door on the front of the house, including a horizontal door jamb;
    - a window on the near side of the house, including a window sill;
    - a chimney on the near side of the roof whose base is parallel to the roof and whose top is horizontal.

    Which vanishing points do you use to draw these objects?

25. Each of the objects you drew in question 24 has several visible faces. (A "face" is a bounded subset of a plane).
    (a) Your drawing should have at least four visible faces that vanish to line ℓ_1. Label those faces.

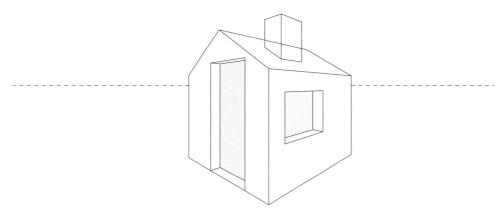

FIGURE 5.1: One possible way to add a door (with door jambs), a window (with a sill), and a chimney to the house.

(b) Your drawing should have at least one visible face that vanishes to line ℓ_2. Label that face (or those faces).

(c) Your drawing should have at least four faces that vanish to line ℓ_3. Label those faces.

(d) Your drawing should have at least two faces that vanish to line ℓ_4. Label those faces.

A COMMENT ON NOTATION

The symbol $\mathbb{R}^3$ is used in two different ways in the mathematical literature. The symbol sometimes denotes the real vector space, $\mathbb{R}^3 = \{(x, y, z) \mid x, y, z \in \mathbb{R}\}$; this is not what we mean here. We are very specifically using the symbol $\mathbb{R}^3$ in the sense of Euclidean space, the nebulous geometry of points and lines and angles and such arising from Euclidean axioms, not its coordinate representation.

Surprisingly, there is no standard notation for extended Euclidean space. Some other symbols that mathematicians use include

- E^{3+}, as in Grünbaum's *Configurations of Points and Lines* [29],
- EE^3, as in Row and Reid's *Geometry, Perspective Drawing, and Mechanisms* [46],
- $E\mathbb{R}^3$, as suggested in Donald Robertson's website on Desargues's theorem [45],
- ES_3, as denoted in Rey Casse's *Projective Geometry: An Introduction* [10].

In these modules, we use $\mathbb{E}^3$ to denote extended Euclidean space—even though this is not a standard notation—partly because there *is* no standard notation, and partly because this choice seems to echo the notation we use for other familiar spaces and sets (such as $\mathbb{N}$, $\mathbb{Z}$, and $\mathbb{C}$, and of course $\mathbb{R}^3$).

In a later module, we'll encounter homogeneous coordinates, which live in $\mathbb{R}P^3$, the standard notation for *real projective space*. The space $\mathbb{R}P^3$ is closely related to extended Euclidean space (with much the same relationship as that between the real vector space and Euclidean space).

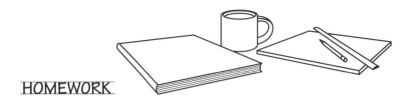

HOMEWORK

EXERCISES

ⓔ 5.1. A *triangle* in $\mathbb{E}^2$ is a set containing three points, non-collinear, and the three lines defined by pairs of those points.
  (a) Is it possible to have a triangle in $\mathbb{E}^2$ with no ordinary points? If so, draw an example; if not, explain why not.
  (b) Is it possible to have a triangle in $\mathbb{E}^2$ with exactly one ordinary point? If so, draw an example; if not, explain why not.
  (c) Is it possible to have a triangle in $\mathbb{E}^2$ with exactly two ordinary points? If so, draw an example; if not, explain why not.
  (d) Is it possible to have a triangle in $\mathbb{E}^2$ with three ordinary points? If so, draw an example; if not, explain why not.

ⓔ 5.2. In Figure 5.0, the lines ℓ_2, ℓ_3, and ℓ_4 intersect at P_3. In general, if the three distinct vanishing lines of three planes α, β, and γ intersect in a common point, what can you say about the lines formed by intersections of those planes? (That is, how do the lines $\alpha \cdot \beta$, $\beta \cdot \gamma$, and $\gamma \cdot \alpha$ relate to one another?)

ⓔ 5.3. In Figure 5.0, the lines ℓ_1, ℓ_2, and ℓ_4 form a triangle (that is, they do not intersect in a common point). In general, if the three distinct vanishing lines of three planes α, β, and γ do not pass through a common point, what can you say about the lines formed by intersections of those planes? (That is, how do the lines $\alpha \cdot \beta$, $\beta \cdot \gamma$, and $\gamma \cdot \alpha$ relate to one another ?)

ⓔ 5.4. (With thanks to the *Emissary* [6]). Let A be a corner of a cube as in Figure 5.2, and B and C be midpoints of their respective edges. Consider the polygon $\mathcal{P}$ that is the intersection of the (solid) cube and the plane containing the triangle ABC. How many sides does $\mathcal{P}$ have?

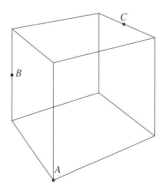

FIGURE 5.2: What is the intersection of the plane *ABC* with this cube?

Draw the polygon $\mathcal{P}$. At each step of the construction, justify your reasoning that the lines/points/planes you are constructing exist. (For example, if you use the intersection of two lines in your construction, explain why this intersection is non-empty).

(E) 5.5. In future modules, having these fundamental facts about $\mathbb{E}^3$ gathered in one place will be helpful to us. Fill in the blanks:

Theorem E1: Two distinct points in $\mathbb{E}^3$ determine a unique _____.

Theorem E2: Two distinct lines in $\mathbb{E}^3$ lie in the same _____ if and only if the lines _____ in exactly one _____.

Theorem E3: Two distinct planes in $\mathbb{E}^3$ determine a unique _____.

Theorem E4: A plane and a line not on the plane determine a unique _____.

Theorem E5: A line and a point not on the line determine a unique _____.

ART ASSIGNMENT

⚠ 5.1. Take a photograph that illustrates several vanishing lines, that is, images of ideal lines.

This photograph is related to Art Assignment 3.2, in which you photographed objects and indicated their vanishing points. In this assignment, you should photograph an object, several objects, or an architectural structure with several planar faces. You should photograph the scene in such a way that the sets of lines in at least three planar directions are clearly visible. Either digitally or by hand, add vanishing points and vanishing lines.

To add the vanishing points, extend the images of parallel lines, using dashed lines of your own, to a vanishing point. To draw vanishing lines, use solid lines to connect relevant vanishing points, or use parallel lines as appropriate. Your embellished photograph should contain at least three vanishing lines. This means you will need to think carefully beforehand about what you photograph, and also about the angle at which you photograph it.

You are likely to discover that you will need to shrink the photograph and/or attach extra paper, because the vanishing points and lines are likely to be off the edges of the picture.

PROOF/COUNTEREXAMPLE

[P] 5.1. Any non-incident line and point in $\mathbb{E}^3$ determine exactly one plane. (A line and point are *incident* if the point is a subset of the line).

[P] 5.2. Any three non-collinear points in $\mathbb{E}^3$ determine exactly one plane.

A Note on Etymology and Spelling

Why do we spell "collinear" with two "l"s? The etymological origins of the word are likely from *con-* (originally from the preposition *cum*, "together," as in *magna cum laude*) and *līneāris* ("of or belonging to lines, consisting of lines"). The word *līnum* meant "flax," a plant that is still commonly referred to by its Latin name *Linum usitatissimum*. Long ago, people commonly made thread, ropes, or fishing lines or nets with flax. Threads and ropes look like lines when they're pulled taut, and so the Latin word for flax eventually gave us *line*.

So, if "collinear" comes from *con* and *līneāris*, why don't we spell this word "*con*linear"?

The reason is that spellings of words change over time through a process called "assimilation" [27], often by the changing (assimilating) of the final consonant of a preposition to the first consonant of the noun, similarly to the way "want to" begins to sound like "wanna" when people speak informally. A more familiar example of the process of assimilation is that when we work (labor) together, we don't *con*laborate, we *collaborate*.

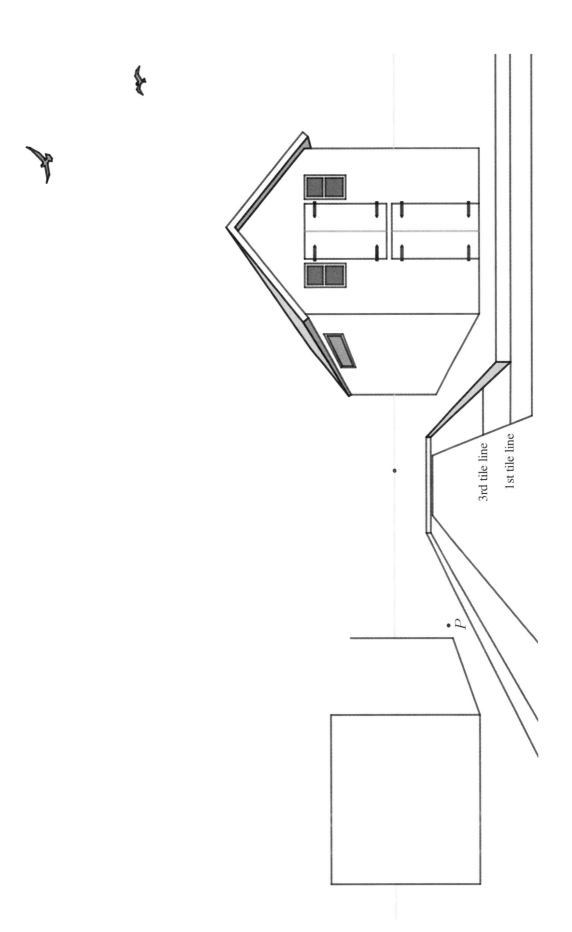

3rd tile line

1st tile line

P

FIGURE 6.0: The start of a sketch in one-point perspective inspired by boat houses in Smögen, Sweden. [For use with the OF MESHES AND MAPS module.]

6

Of Meshes and Maps

Revisiting the image of a line

Overview In this module, we will define what we mean by a "projection" from certain subsets of $\mathbb{E}^3$ onto an image plane. We first describe the objects that we will make images of; that is, we'll define a "mesh." From there, we define a function called a *mesh map*, and compare and contrast this mesh map with a plane-to-plane projection called a *perspective collineation*. Finally, we uses meshes and their images to duplicate and divide shapes in perspective.

In much of mathematics, we think of points as the fundamental objects in space, and so we define lines as a special kind of collection of points. But when we draw a perspective picture, it's clear that lines are fundamental objects in their own right. After all, we don't draw every single little point along that line; we just draw the line.

Moreover, when we draw a picture, we don't draw every single point and every single line that exists in the world; we draw a special subset of points and lines that seem to be important to us: the edges and corners of the house, the corners and edges of the 'T', etc. The following definition gets at the notion of a drawable object that has a set of points and lines that relate to one another in a natural way.

Definition A *mesh* $\mathcal{M} = \{P_1, \ldots, P_k, \ell_1, \ldots, \ell_n\}$ is a collection of points and lines in $\mathbb{E}^3$ such that every point in $\mathcal{M}$ is incident with at least two lines in $\mathcal{M}$ and every line in $\mathcal{M}$ is incident with at least two points in $\mathcal{M}$.

Note that points are *elements* of $\mathbb{E}^3$ and lines are *subsets* of $\mathbb{E}^3$, so we can not properly say $\mathcal{M} \subset \mathbb{E}^3$. Nonetheless, we will occasionally abuse notation and say it anyway.

1. Figure 6.1 shows three collections of points and lines: the word "Hi," a star, and a sketch of a house. Which of these collections (if any) forms a mesh?

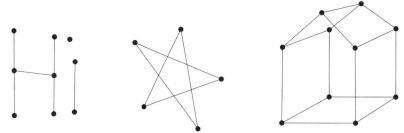

FIGURE 6.1: Three collections of points and lines: the word "Hi," a star, and a sketch of a house. (For mathematical reasons, we should draw the whole lines, but to make the pictures easier to understand, we draw only the line segments; think of these as "abbreviations" that stand for the entire line.)

2. If k and ℓ are lines in a mesh $\mathcal{M}$, must $k \cdot \ell$ be an element of $\mathcal{M}$?

For the questions 3–9 below, determine if it is possible to create the specified meshes. If such a mesh is possible, draw an example; if it is not possible, give a brief explanation as to why not.

3. A mesh with exactly one point?

4. A mesh with exactly two points?

5. A mesh with exactly three points?

6. A mesh with exactly three points that does not create a triangle?

7. A mesh with exactly four points?

8. A mesh with fewer points than lines?

9. A mesh with fewer lines than points?

WHAT IS AN IMAGE?

In our IMAGE OF A LINE module, we pointed out, "We have not yet defined exactly what we mean by 'the projection via O onto the picture plane ω',' and …there are several different things we might mean by this phrase." Here we get rid of that ambiguity and formally define what we mean by a projection in the context of meshes.

> **Definition** Given a mesh $\mathcal{M}$ in $\mathbb{E}^3$, a plane $\omega' \subset \mathbb{E}^3$, and a point $O \in \mathbb{E}^3$ not incident with $\mathcal{M}$ or ω', we can define a *mesh map* as
>
> $$\prime = \mathcal{M} \to \mathcal{M}' \subset \omega'$$
>
> with center O, taking points to points and lines to lines such that
>
> - O, P, P' are collinear for every point $P \in \mathcal{M}$ and
> - O, ℓ, ℓ' are coplanar for every line $\ell \in \mathcal{M}$.

In the definition of mesh map, we require that O be incident with neither $\mathcal{M}$ nor ω'.

10. Add the labels O, ω', P, P', ℓ, and ℓ' to Figure 6.2 as appropriate.
11. Why is the mesh map not well defined if O is a point in $\mathcal{M}$?

12. Why is the mesh map not well defined if O lies on a line ℓ in $\mathcal{M}$?

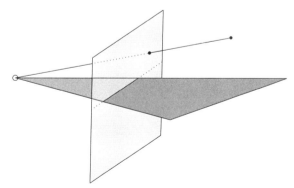

FIGURE 6.2: In this image, we show a single point P and single line ℓ from some larger mesh $\mathcal{M}$. Add the labels O, ω', P, P', ℓ, and ℓ' that illustrate the mesh map. (See problem 10.)

13. Why do we require that $O \notin \omega'$? What goes wrong with the mesh map if O lies in the plane ω'?

In questions 14–19, we assume we have a mesh map $\prime = \mathcal{M} \to \mathcal{M}'$ as described in the definition.

14. [T/F] If P and S are distinct points in $\mathcal{M}$, then P' and S' are distinct.

["Distinct" means "not equal".]

15. [T/F] If $Q \in \ell$ and if both Q and ℓ are in $\mathcal{M}$, then $Q' \in \ell'$.

16. [T/F] If P and Q are distinct points in $\mathcal{M}$ with PQ also in $\mathcal{M}$, then $P'Q' = (PQ)'$. That is, the image of the line between two points is the line between the two images of the points.

17. [T/F] If P and Q are distinct points in $\mathcal{M}$ and $P'Q' \in \mathcal{M}'$, then $PQ \in \mathcal{M}$.

18. [T/F] If k and ℓ are distinct lines in $\mathcal{M}$ with $k \cdot \ell$ also in $\mathcal{M}$, then $k' \cdot \ell' = (k \cdot \ell)'$. That is, the image of the intersection of two lines is the intersection of the images of the lines.

19. [T/F] If k and ℓ are distinct lines in $\mathcal{M}$ and $k' \cdot \ell' \in \mathcal{M}'$, then $k \cdot \ell \in \mathcal{M}$.

LOOKING AHEAD: COLLINEATIONS

In a future module, we will explore the special case where the mesh $\mathcal{M}$ is planar (for example, the lines in a fence or a sidewalk, rather than in a cube or a building). Let us consider how this assumption changes our answers to the questions above.

20. For the statements below, assume we have planar meshes $\mathcal{M} \subset \omega$ and $\mathcal{M}' \subset \omega'$, a point O not in ω or ω', and a mesh map $\prime = \mathcal{M} \to \mathcal{M}'$ as described in the definition.

    (a) [T/F] If P and S are distinct points in $\mathcal{M}$, then P' and S' are also distinct.

    (b) [T/F] If $Q \in \ell$ and both Q and ℓ are in $\mathcal{M}$, then $Q' \in \ell'$.

    (c) [T/F] If P and Q are distinct points in $\mathcal{M}$ with PQ also in $\mathcal{M}$, then $P'Q' = (PQ)'$. That is, the image of the line between two points is the line between the two images of the points.

    (d) [T/F] If P and Q are distinct points in $\mathcal{M}$ and $P'Q' \in \mathcal{M}'$, then $PQ \in \mathcal{M}$.

(e) [T/F] If k and ℓ are distinct lines in $\mathcal{M}$ with $k \cdot \ell$ also in $\mathcal{M}$, then $k' \cdot \ell' = (k \cdot \ell)'$. That is, the image of the intersection of two lines is the intersection of the images of the lines.

(f) [T/F] If k and ℓ are distinct lines in $\mathcal{M}$ and $k' \cdot \ell' \in \mathcal{M}'$, then $k \cdot \ell \in \mathcal{M}$.

Mesh maps give us a way to mathematically describe what we do naturally when we draw a three-dimensional object in perspective—we look at only a select few points and lines and project them onto a plane through a center point. However, as we saw in the questions 14–19 above, a general mesh map is not necessarily invertible and can give us images with different incidence properties than those of the original mesh.

Question 20 shows that if we restrict our mesh to a plane, and make sure the center O lies off of both the object and image plane, the mapping becomes much nicer. The "nice" comes from the fact that the mapping becomes a one-to-one correspondence, with each point or line in $\mathcal{M}$ corresponding to a unique point or line in $\mathcal{M}'$ and vice versa. It is always possible to extend such a mapping to the entire plane, so that $': \omega \to \omega'$, in which case we call it a *perspective collineation*. We will explore this special class of maps further in Chapter 8.

APPLICATIONS TO DRAWING

Let us put theory into practice by solving some drawing puzzles. Figure 6.0 shows the beginning of a sketch of three houses and a sidewalk in one-point perspective. The house on the right is almost finished; the first house on the left has some walls but still needs a roof; the second house on the left has only one corner so far (the near bottom corner at point P). The next set of questions will lead you to finishing the sketch.

The face containing the peak of the house on the right faces perpendicularly to the sidewalk. For the two houses on the left, we will want the peak to be parallel to the sidewalk. This means we will need to figure out how to determine where to draw the middle of the house—a potentially hard question!

21. Let us start with a simpler question, by starting on the ground. We have drawn the first and third sidewalk lines. How do we divide this image of a rectangle in "half" so that we get the second sidewalk line?

This question is not easy, and there are multiple correct solutions. Take your time and try to find one or more constructions.

22. Use your favorite construction from question 21 to
(a) locate a vertical line containing the peak of the roof of the house on the left, and
(b) finish drawing the visible edges of that house.
(*Hint*: You can choose to make the roof as tall as you like, but to make it possible to draw the entire roof, you might want to make the roof shallow enough so that the vanishing point of the edge of the roof is on the paper.)

23. (a) For both of the previous questions, we are implicitly using two important properties of mesh maps:
      - The image of the intersection of two lines is the _____ of the images of those lines, and
      - the image of the line between two points is the line between the _____ of those points.

    (b) In your construction for question 21, explain what the mesh $\mathcal{M}$ represents.

    (c) In your construction for question 21, explain what the plane ω' represents.

    (d) In your construction for question 21, explain what the mesh $\mathcal{M}'$ represents.

    (e) Does the vanishing point have a preimage under this mapping? In other words, is there a point in the mesh $\mathcal{M}$ that maps to V?

24. Suppose the house on the right is a glass house, so that all four edges of the floor, as well as all edges and corners of the house and roof, are visible.

    (a) Draw these hidden edges and points that represent the three-dimensional mesh of the house.

    (b) Why is this image of a mesh map problematic, from a perspective drawing point of view?

Now that we have successfully divided objects in half, let us duplicate objects.

25. Continue the sidewalk into the distance. You already have the first three sidewalk lines; draw the fourth sidewalk line.
    (Again, there are many correct ways to do this.)

26. Draw several more sidewalk lines, continuing the sidewalk tiles into the distance.

27. Draw the second house on the left so that it is the image of one identical to the first house on the left, with the near corner at the point P.

The next question asks for a more difficult division than that in questions 21 and 22.

28. Draw a door for the house on the right in the wall that faces the sidewalk. You may make the door any height you want, but the door should be one third of the width of the wall, and horizontally in the center of that wall. That is, you will need to determine a way to divide the wall into three equal parts (in a perspective sense).

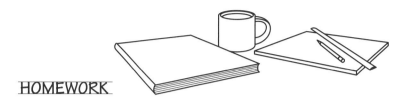

HOMEWORK

EXERCISES

(E) 6.1. Is it possible to create a mesh that has
   (a) exactly one line?
   (b) exactly two lines?
   (c) exactly three lines?
   (d) exactly four lines?

(E) 6.2. Draw a mesh with exactly five lines that has
   (a) exactly four points.
   (b) exactly five points.
   (c) exactly six points.
   (d) exactly ten points.

   If any of the above is impossible, explain your reasoning.

(E) 6.3. A triangle is an important example of a mesh, as we shall see in the Desargues's theorem chapter. Two other important examples of meshes are the *quadrangle* and the *complete quadrangle*, as defined below.
   (a) A *quadrangle* is a set of four coplanar points, such that no three are collinear, as well as the n lines defined by pairs of these four points. What is the value of n?
   (b) Explain why a quadrangle is a mesh using the definition.
   (c) A *complete quadrangle* is a quadrangle, plus the other three points of intersection defined by the n lines. These three points are called the *diagonal points*. Explain why a complete quadrangle is a mesh using the definition.

   The exercises below are a review (and possibly an expansion) of the questions you worked on in class.

(E) 6.4. Figure 6.3 shows the beginning of four different constructions that allow us to divide a fence panel in half. In each construction, add either one or two more construction lines and then the desired halfway line.
   (a)–(c) For Figures 6.3(a)–(c), draw the corresponding construction on a rectangle that explains why the construction is valid.
   (d) For Figure 6.3(d), draw a plan view explaining why this construction works.
   (e) Which of these constructions would still work if the fence panel were in two-point perspective instead of one-point perspective?

(E) 6.5. How would you modify the constructions in Figure 6.3 to divide a fence panel into three equal vertical pieces?

(E) 6.6. How would you modify the constructions in Figure 6.3 to double the fence panel?

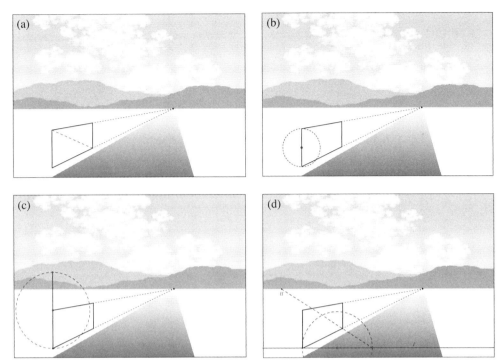

FIGURE 6.3: The start of four different constructions to divide a fence panel in half, perspectively (see homework exercise Ⓔ6.4). The circles indicate equal distances between the center and indicated points on the circumference. In (d), *H* is an arbitrary point on the horizon and *ℓ* is a line parallel to the horizon.

ART ASSIGNMENT

⚠ 6.1. Take two photographs of something in the real world that contains regularly spaced objects (or of something that is divided into a number of equal pieces). Photograph this object from two locations, in a way such that both of these pictures are in one-point perspective. Print your pictures—you might want to lighten the images first, because you will be drawing on top of them. See Figure 6.4 for an example.

In one picture, the face with the repeated objects should be parallel to the picture plane, that is, they should all be the same distance from the picture plane. Verify that the first image contains collections of parallel, evenly spaced segments. Measure those segments in the photo, and write the measurements on your picture.

In the second picture, the face with the repeated objects should be perpendicular to the picture plane, and the repeated object should get further away from the camera. Draw two different constructions onto this photo: one construction should be based on intersections of diagonal lines, and another should be based on vanishing points of diagonal lines. Both of these sets of lines you draw should verify that the fencepost construction techniques we used in class work.

FIGURE 6.4: Two photographs by Biyang Sun, F&M Class of '16. The first photo shows that the fence along the track contains evenly spaced posts; the "×" and lines to the vertical horizon that we added in the second photo show that our construction techniques work. The lines of the track lanes helped Biyang take the second picture in one-point perspective, because they gave her something that was *not* the fence, but runs parallel to the fence, to aim at.
Courtesy of Biyang Sun

PROOF/COUNTEREXAMPLE

A *triangle* (a mesh with exactly three points, or equivalently, a mesh with exactly three lines) is an example of a *full mesh*; that is, a mesh satisfying

- for every pair of distinct points P and Q in the mesh, the line PQ is also in the mesh, and
- for every pair of distinct lines ℓ and k in the mesh, the point $\ell \cdot k$ is also in the mesh.

Use this definition of *full mesh* to prove or to find a counterexample for statements 6.1 and 6.2 below.

[P] 6.1. If a full mesh $\mathcal{M}$ is not empty, then $\mathcal{M}$ contains at least six elements.

[P] 6.2. If a full mesh $\mathcal{M}$ contains at least four points, then $\mathcal{M}$ contains infinitely many elements.

[P] 6.3. Suppose a mesh has n points ($n \geq 3$), no three of which are collinear. Create a conjecture about the smallest possible number of lines in the mesh, and prove your conjecture.

[P] 6.4. Suppose a mesh has n points ($n \geq 3$), no three of which are collinear. Create a conjecture about the largest possible number of lines in the mesh, and prove your conjecture.

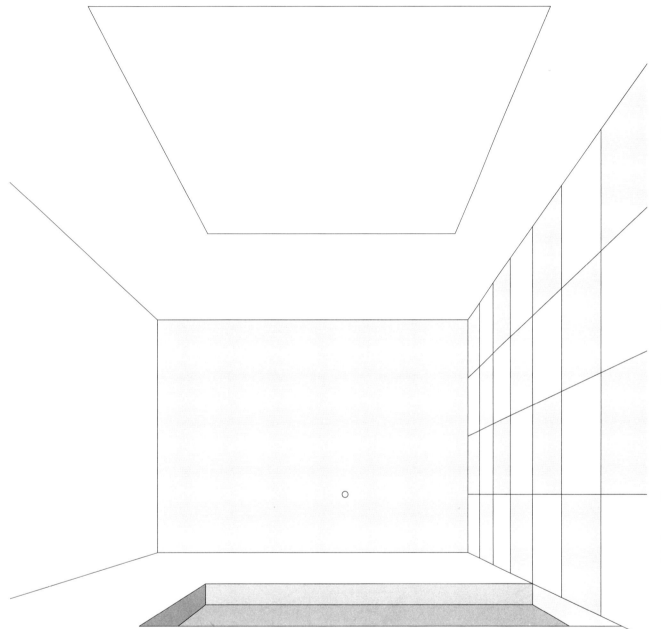

FIGURE 6.1.0: A hallway with space for a poster. [For use with the FIELD TRIP: PERSPECTIVE POSTER module.]

3. Draw vertical and horizontal lines (or lines to the vanishing point) on each image.
4. Then subdivide again with more ×'s.

5. Further subdivide one section at a time, until you feel confident you can transfer the image of the cartoon onto your poster.

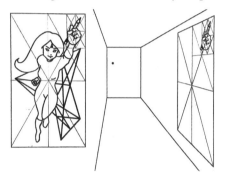

6. Some students find it helpful to number the cells of the grid. It's easy to get lost!
7. When you've transferred the image, erase the construction lines and admire your work!

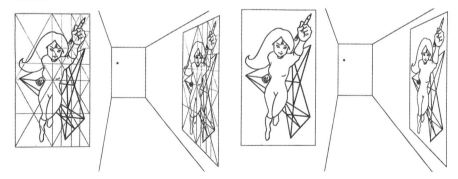

You're a superhero!

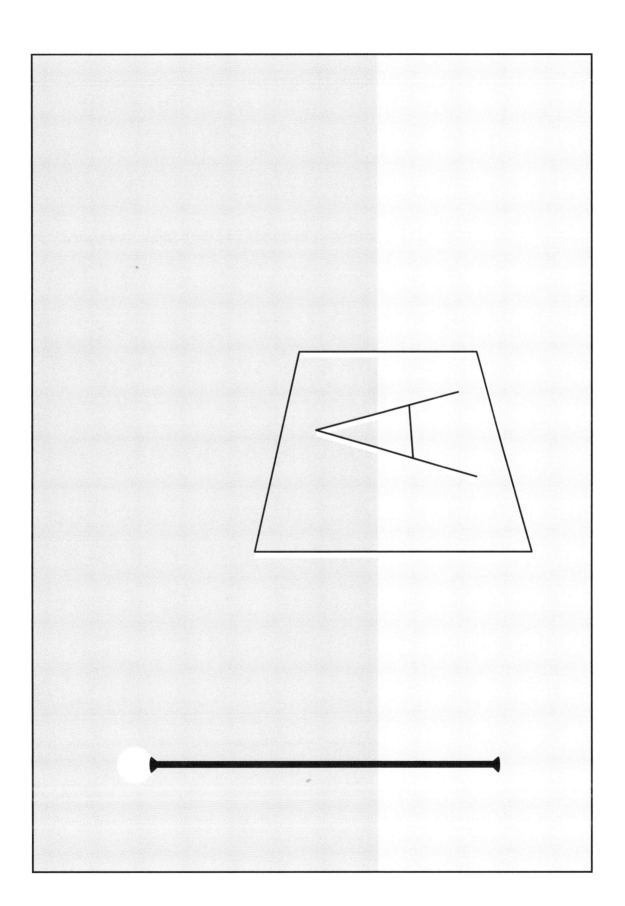

FIGURE 7.1.0: A streetlamp shining on a vertical "A." [For use with the EXPLORATION AND DISCOVERY module.]

7

Desargues's Theorem

7.1 Exploration and Discovery

Overview In this module, we will discover and explore one of the most useful theorems in all of projective geometry: Desargues's theorem! To motivate this theorem, we introduce an activity we call "Making an A in math class!"

Consider the drawing in Figure 7.1.0; it shows a vertical lamppost and a vertical pane of glass etched with the letter "A." The light at the top of the lamppost should cast shadows of all of these objects onto the horizontal plane of the ground. Our goal will be to draw the shadows that ought to be in this picture.

1. First, describe the shadow of the lamppost itself. Is the shadow of the lamppost a point? A line segment? A line? A plane? Where is the shadow?
2. Next, let us determine the shadow of the near vertical edge of the glass. Because the glass is resting on the ground, we know the shadow will meet the edge of the glass at that point. But in which direction does the shadow go (see Figure 7.1)?

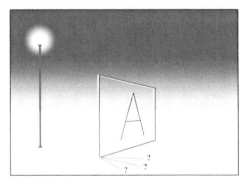

FIGURE 7.1: How do we draw the shadow of the edge of the glass? See question 2.

3. How do we know that the shadow of this edge runs off the page, that is, how do we know the top near corner of the glass has a shadow that is not on the paper?

GEOGEBRA exploration

Now fire up GEOGEBRA on a computer; turn off the axes and labeling. The goal of the next few steps is to construct something like the drawing in Figure 7.4.

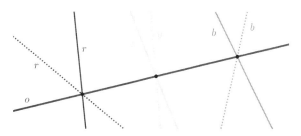

FIGURE 7.4: A GeoGebra set-up for exploring triangles perspective from a line.

10. Draw an axis o (a line between two points).
11. At one of those points, add two more lines. You might color those lines red; Figure 7.4 labels these as "r." **Important:** Make sure that you construct these lines so that even if you use the arrow key to move the lines around, these two red lines are always concurrent with o; that is, the three lines always intersect at a common point.
12. At the other of those points, add two more lines. You might color those lines blue; Figure 7.4 labels these as "b." Again, the two blue lines and o should always be concurrent.
13. Add a third point to the axis o, and add two lines (yellow or "y") concurrent with o.
14. Finally, to distinguish which lines belong to which mesh, change one line associated with each point to a dotted line. (It doesn't matter which one, since you are able to move the lines around).

At this point, we have two sets of lines that are perspective from the line o. To turn each set of lines into a mesh, we need to add points.

15. Locate the intersection point of the solid red and yellow lines; color this point orange. Do the same for the dotted red and yellow lines. Below, we will refer to these points as "N" for "oraNge".
16. Locate the intersection point of the solid red and blue lines; color this point purple. Do the same for the dotted red and blue lines. Below, we will refer to these points as "P".
17. Locate the intersection point of the solid yellow and blue lines; color this point green. Do the same for the dotted yellow and blue lines. Below, we will refer to these points as "G".
18. Now you have two meshes, each of the form $\{N, P, G, r, b, y\}$. (It may be easier to see what you are doing if you use the polygon tool to define and shade the interior of these two triangles.) Are these meshes perspective from a line? How do you know?
19. Is it possible to move your lines and points around so that these meshes are perspective from a point as well? Connect the orange points, the purple points, and the green points to explore this question.

20. Based on your experimentation above, make your best guess as to how to finish the sentence below.

Conjecture 7.1. *If two triangles are perspective from a line, then ....*

I freely confess that I never had taste for study or research either in physics or geometry except insofar as they could serve as a means of arriving at some sort of knowledge of the proximate causes for the good and convenience of life, in maintaining health, in the practice of some art ... having observed that a good part of the arts is based on geometry ... that of perspective in particular. –Girard Desargues [33]

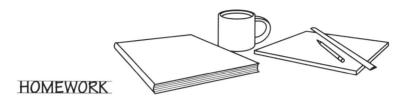

HOMEWORK

EXERCISES

Ⓔ 7.1.1. How would the shadows change if the light came not from a lamppost, but rather from the sun, far off in the distance? (That is, we imagine that the sun is not overhead, but instead is rising or setting).

Ⓔ 7.1.2. Figure 7.5 shows a light source, a kite, the line where the plane of the kite intersects the plane of the ground, and the shadow of one corner of the kite. Draw the remainder of the shadow.

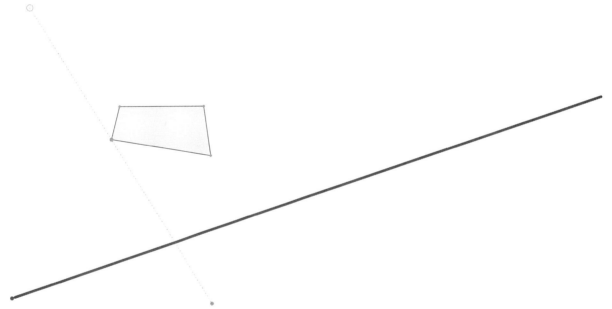

FIGURE 7.5: Draw the shadow of the kite.

THREE CONSTRUCTIONS: A HISTORICAL DIGRESSION

The next three exercises compare three constructions—two of them correct, and one incorrect—for drawing shapes in perspective. The motivation for these exercises requires a bit of historical background first.

The inset in the top left of Figure 7.6 illustrates a classic problem and solution in perspective drawing. Namely, given a figure—the letter F in this case—on the front face of a cube drawn in one-point perspective, draw its reflected image on the top face, in correct perspective. In his book *De Prospectiva pingendi*, or *On the Perspective of Painting* [44], the Renaissance painter Piero della Francesca gave a solution of this problem, which he applied to the case of a regular pentagon as shown in the lower left of Figure 7.6. This is a reproduction of Figure 75 from Dan Pedoe's *Geometry and the Visual Arts* [41]. The right

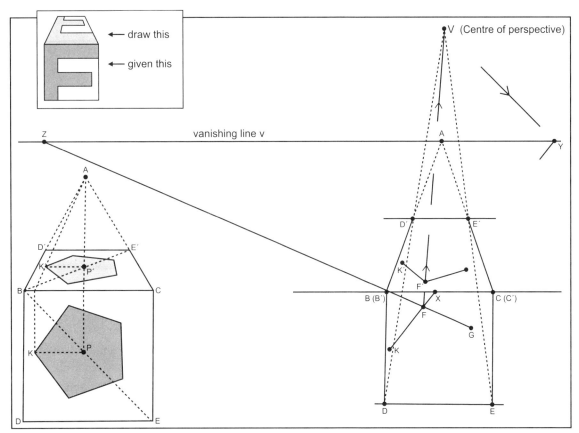

FIGURE 7.6: A simplified drawing of Figure 75 from Dan Pedoe's *Geometry and the Visual Arts* [41].

side of the figure shows Pedoe's alternate solution applied to two sides of the pentagon. Pedoe only deals with two of the pentagon edges, *KF* and *FG*, assuming that the reader can take it from there.

Pedoe writes,[1]

> Pierro della Francesca, in a book published between 1470 and 1490 showed how this could be done. We shall obtain identical results by using Desargues Theorem …
>
> If *KF* meets the axis in the point *X*, then *K′F′* passes through *X*. If *KF* meets the vanishing line *v* in *Y*, then the line *K′F′* is parallel to *VY*. So, to obtain *K′*, which lies on *VK*, we find the intersection of *VK* and the line through *X* parallel to *VY*. To find *F′*, we obtain the intersection of *VF* and the line through *X* which [*sic*] is parallel to *VY*, and similarly for the points *G′*, *H′* and *I′*. This gives a rapid method for constructing plane perspective drawings. [41, pp. 186–189]

Surprisingly, Pedoe's solution is wrong. (Moreover, the solution does not make use of Desargues's theorem.) Despite the fact that the error in Pedoe's construction is easy to detect, it has survived several editions and publishers, including St. Martin's Press (1976), Penguin Books (1976), General Publishing (1976), and Dover (1983 and 2011). Indeed,

---

[1] We reproduce Pedoe's text with its original spelling (e.g., *Pierro* vs. *Piero*) and punctuation.

ART ASSIGNMENT

⚠ 7.1.1. In GeoGebra, draw a perspective picture with the following objects:
- a horizon line with appropriate vanishing points indicated;
- two three-dimensional objects (such as cubes or rectangular boxes or even letters) sitting on the ground, drawn in one-point perspective:
  - a cube (indicate the viewing distance with a "diagonal" vanishing point), and
  - a challenge to yourself—for example, a house or a letter;
- a lamppost (that is, a point light source with a vertical* post and a visible base where the post meets the ground); and
- the shadows that your objects cast on the ground because of this lamp. In particular, you should show the shadows of every point and of every line segment in your mesh. Each face of your object (which should be a polygon) should have its own polygon as a shadow.

Hide all construction lines, but leave the vanishing points visible.

Your picture should look "nice."

You should submit three versions of this project:
- A paper version with all relevant objects (objects, shadows, horizon line, vanishing points) visible, but the construction lines hidden.
- Another paper version, with the objects moved around, but still all visible as before.
- A GeoGebra version. If your name is Sam Smith, then name your file "PG-shadow-Sam-Smith.ggb."

  *Because you create this drawing in GeoGebra, you should make sure that all "vertical" line segments remain perpendicular to the horizon even when you move points around!

PROOF/COUNTEREXAMPLE

P 7.1.1. [T/F] Given six distinct points A_1, B_1, C_1, A_2, B_2, and C_2 in $\mathbb{E}^3$, let

$$\mathcal{S}_1 = \{A_1, B_1, C_1, A_1B_1, C_1A_1, B_1C_1\};$$

define $\mathcal{S}_2$ accordingly. Suppose moreover that $\mathcal{S}_1$ and $\mathcal{S}_2$ are perspective from exactly one point. Then they are perspective from exactly one line.

   [*Note*: When we prove Desargues's theorem in class, we will assume more structure than we do in this homework exercise; that is, we will require three points, *and* three lines, *and* that they form a mesh. But in these homework exercises, we do not assume either $\mathcal{S}_1$ or $\mathcal{S}_2$ is a mesh.]

P 7.1.2. [T/F] Given six distinct lines j_1, k_1, l_1, j_2, k_2, and l_2 in $\mathbb{E}^3$, let $\mathcal{S}_1$ be the set $\{j_1, k_1, l_1, j_1k_1, l_1j_1, k_1l_1\}$. Define $\mathcal{S}_2$ accordingly. Suppose moreover that $\mathcal{S}_1$ and $\mathcal{S}_2$ are perspective from exactly one line. Then they are perspective from exactly one point.

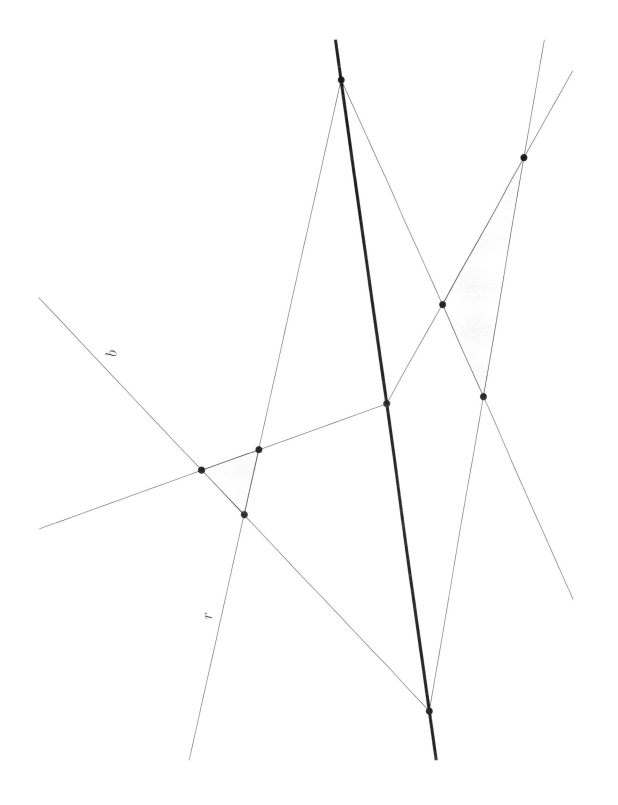

FIGURE 7.2.0: A setup to help us visualize the proof of Desargues's theorem. [For use with the WORKING TOWARD A PROOF module.]

7.2

Working toward a Proof

Overview In this module, we will prove the theorem you conjectured in the last module!

Desargues's Theorem and Its Converse. *Two triangles in $\mathbb{E}^3$ are perspective from a line if and only if they are perspective from a point.*

We will prove this theorem in four steps:

1. Triangles in different planes that are perspective from a point are perspective from a line.
2. Triangles in different planes that are perspective from a line are perspective from a point.
3. Triangles in the same plane that are perspective from a line are perspective from a point.
4. Triangles in the same plane that are perspective from a point are perspective from a line.
   (You will complete the last step as a homework assignment).

Before we begin with the proof below, let us recall five fundamental facts about $\mathbb{E}^3$. Fill in the following blanks:

Theorem E1: Two distinct points in $\mathbb{E}^3$ determine a unique _____.

Theorem E2: Two distinct lines in $\mathbb{E}^3$ lie in the same _____ if and only if the lines _____ in exactly one _____.

Theorem E3: Two distinct planes in $\mathbb{E}^3$ determine a unique _____.

Theorem E4: A plane and a line not on the plane determine a unique _____.

Theorem E5: A line and a point not on the line determine a unique _____.

STEP 1. TRIANGLES IN DIFFERENT PLANES THAT ARE PERSPECTIVE FROM A POINT ARE PERSPECTIVE FROM A LINE.

Assume two triangles $\mathcal{T} = \{N, P, G, r, b, y\}$ and $\mathcal{T}' = \{N', P', G', r', b', y'\}$, in perspective from a point O, lie in distinct planes ω and ω' in $\mathbb{E}^3$. To make the proof a tiny bit easier,

13. *GG′* is [the empty set / a point / a line / a plane / we can't determine this].

14. In which of the above plane(s)—if they exist—does *GG′* lie?

15. [T/F] *GG′* and *NN′* must lie in a common plane. (If so, which plane(s)? If not, why not?)

16. [T/F] $(GG′) \cdot (NN′)$ is [the empty set / a point / a line / a plane / we can't determine this].

17. The point $(GG′) \cdot (NN′)$ — if it exists — lies in which plane(s)? Does it lie in ω? $\omega′$? ω_{yellow}? ω_{red}? ω_{blue}?

18. [T/F] The line *PP′* lies in the plane ω_{yellow}.

19. $(PP') \cdot \omega_{yellow}$ is [the empty set / a point / a line / a plane / we can't determine this].

20. [T/F] The three planes ω_{yellow}, ω_{red}, and ω_{blue} meet in a single point.

21. Therefore, the triangles $\mathcal{T}$ and $\mathcal{T}'$ are perspective from the point ____. ■

22. Why does this set of questions and answers fail to prove the statement in question 21 for two triangles in the *same* plane? Which of the above statements/questions would have a different answer in that situation?

The third step in the proof of Desargues's theorem follows from, or else allows us to prove, the following proposition; we leave the proof of the proposition as a homework exercise.

Perspective Proposition Suppose two meshes $\mathcal{M}$ and $\mathcal{N}$ in $\mathbb{E}^3$ are perspective from some point P. Then we can create a new, larger mesh

$$\mathcal{P} = \mathcal{M} \cup \mathcal{N} \cup \{P\} \cup \{\text{lines of perspective}\}$$

by taking the union of our original meshes together with the point P and the lines that connect P to the points of $\mathcal{M}$ and $\mathcal{N}$. Choose some point O not incident with $\mathcal{P}$ and some plane ω' not incident with O. This gives us a mesh map $\iota : \mathcal{P} \to \omega'$. Then the images $\mathcal{M}'$ and $\mathcal{N}'$ are perspective from a point in the plane ω'.

23. Consider Figure 7.10. If you think about the house and its shadow, both as subsets of $\mathbb{E}^3$, are they perspective from a point? Are the 3-D house and its shadow perspective from a line?

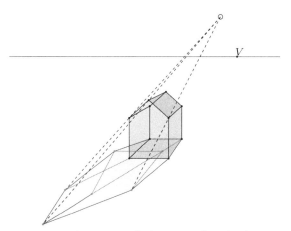

FIGURE 7.10: The image of a house and its shadow, for questions 23–25. The edges of the house are solid and cast solid shadows; the walls are translucent and let some light through.

24. What about the images on this paper? Are the images of the house and its shadow perspective from a point? Are they perspective from a line?

25. Think about Figure 7.10 as a partial illustration of the Perspective Proposition. What elements of this situation might correspond to

- $\mathcal{M}$?
- $\mathcal{N}$?
- P?
- {lines of perspectivity}?

- $\mathcal{M}'$?
- $\mathcal{N}'$?
- P'?
- {lines of perspectivity}$'$?
- O?

STEP 3. TRIANGLES IN THE SAME PLANE THAT ARE PERSPECTIVE FROM A LINE ARE PERSPECTIVE FROM A POINT.

Use the Perspective Proposition, together with the statement 21 from above, to complete the following questions. We begin now with two triangles $\mathcal{T}$ and $\mathcal{S}$, both lying in some plane $\omega \subset \mathbb{E}^3$. Suppose $\mathcal{T}$ and $\mathcal{S}$ are perspective from a line o.

26. For our proof, we'll want to draw a triangle $\mathcal{U}$ in another plane α, so that $\mathcal{T}$ and $\mathcal{U}$ are perspective from a line. How might we want to choose this new plane α? (*There are lots of possible answers to this question, of course. If you don't like your answers to subsequent questions, come back and redefine α so that the next few answers are easier.*)

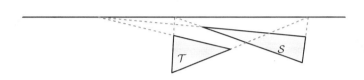

. O

ω

FIGURE 7.11: A figure to help us visualize Desargues's theorem for coplanar triangles.

27. Use your answer to question 26 to draw a plane α in Figure 7.11 above. Here's a drawing suggestion: the picture will probably look better if you draw one set of edges of α *perpendicular* to a set of edges in ω, and draw the other set of edges of α *parallel* to the other set of edges in ω.

28. Having chosen the plane α, consider a point O that is in neither α nor ω. The choice of O defines a mesh map $\prime : \alpha \cup \omega \to \omega$. Let $\mathcal{U}$ be the triangle in α that is perspective with $\mathcal{T}$ from the point O. Therefore $\mathcal{U}' = $ _____.

29. Draw a triangle $\mathcal{U}$ in Figure 7.11. Doing so is not easy; make sure you can explain your steps.

30. What is the name of the line that $\mathcal{U}$ and $\mathcal{T}$ are perspective from? (Your answer here depends on your answer to question 26 above).

31. If we chose α correctly in question 26, then $\mathcal{U}$ should *also* be perspective with $\mathcal{S}$ from a line. How do we know this is so?

32. Which statement above tells us that, therefore, $\mathcal{S}$ and $\mathcal{U}$ are perspective from a point?

33. Where is/what is $\mathcal{S}'$?

34. Put the pieces together. We know $\mathcal{S}$ and $\mathcal{U}$ are perspective from a point. Why does this imply the very thing we want to prove: that $\mathcal{S}$ and $\mathcal{T}$ are perspective from a point?

WHERE DID THIS THEOREM COME FROM?

It was in approximately 1640 that the French mathematician and artist Girard Desargues first wrote the theorem that bears his name to this day. Unfortunately, his *Leçon de Ténèbres* has been lost along with many of his other published mathematical and artistic pamphlets. The closest surviving relative comes in the final several pages of a book by one of Desargues's acolytes: Bosse's *La Perspective de Mr Desargues* [8], published a dozen years later, which pulls together both the artistic and the mathematical work of Desargues. But Bosse's diagram is a little confusing: see Figure 7.12.

35. Bosse claims that triangles *DEK* and *abl* are perspective from a point and from a line. What point? What line?

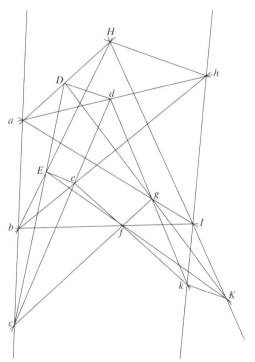

FIGURE 7.12: Bosse's diagram, showing the oldest known description of Desargues's theorem.

36. Shade or color this diagram so that the triangles *DEK* and *abl* appear to lie in different planes in three-dimensional space. Also highlight the lines that connect the triangles to the center. If there are parts that "overlap," color the parts we can see, but not the parts that are blocked from our sight. What can we say about where these triangles lie in relationship to each other? (Notice a lovely symmetry here: we have often noted in this class that *mathematics* helps us understand *art*. But by shading this diagram, you are using *art* to help you see the *mathematics*!)

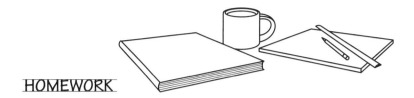

HOMEWORK

EXERCISES

ⓔ 7.2.1. When we proved that two triangles in different planes that are perspective from a line are also perspective from a point, we assumed that the the six points were distinct: that is, $N \neq N'$, $P \neq P'$, and $G \neq G'$. Where did we use this assumption? That is, which of the true statements in questions 10–20 would have different answers if some of the points were not distinct?

ⓔ 7.2.2. Which of the true statements in questions 10–20 depend upon the six lines being distinct: that is, where did we use that $r \neq r'$, $b \neq b'$, and $y \neq y'$?

ART ASSIGNMENT

⚠ 7.2.1. Photograph a real-world instance of Desargues's theorem.

You might choose to stage this photograph yourself (with a lamp, a figure taped to a stick stuck in a mug, and your desk), or you might find an instance while you're walking around outside (the sun making the shadow of a street sign).

In either case, your photograph should contain all of these elements:
- A planar object (possibly a triangle, but other polygonal shapes are acceptable);
- A perspective planar image of that object (such as its shadow on the flat ground); and
- the center of that perspective projection.

You'll want to make sure that the axis of the projection is located in a place where you can draw it by hand on (or near) your photograph; see the explanation below.

You should turn in three versions of this photograph.
- One version should be plain, with no extra lines.
- A second version that is identical, except that you draw (digitally or by hand) the lines from the center through corresponding points of the object and its image.
- A third version that is identical to the first, except that you extend the edges of the object and its image, and show that they meet along an axis. (For this version, depending on the angle of your photograph, you might need to tape the photograph to another piece of paper.)

PROOF/COUNTEREXAMPLE

P 7.2.1. Prove the Perspective Proposition.

P 7.2.2. Prove Step 4: Triangles in the same plane that are perspective from a point are perspective from a line. [*Hint*: Consider the triangles formed by the points *YNN′* and *BPP′*.]

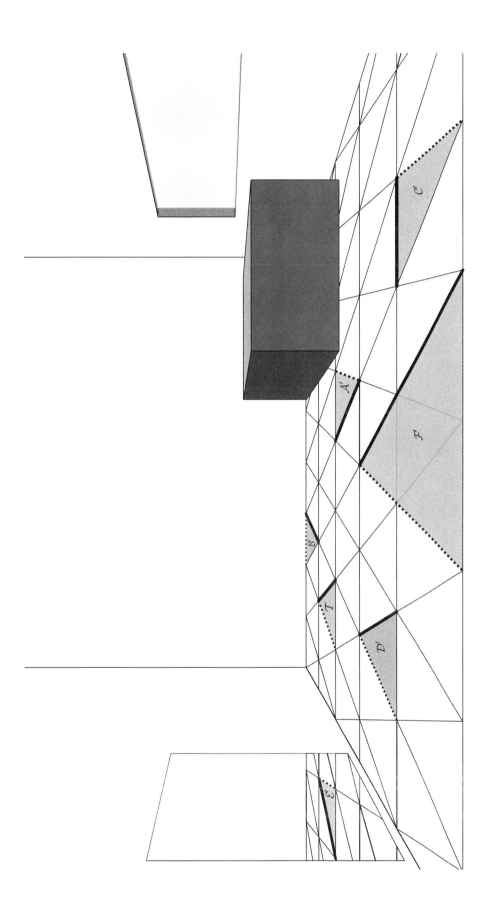

FIGURE 8.1.0: A room with a floor tiled by equilateral triangles. [For use with the HOW PROJECTIVE GEOMETRY FUNCTIONS module.]

8

Collineations

8.1 How Projective Geometry Functions

Overview After our having learned many of the "nouns" of projective geometry (objects, images, points, planes, etc.), this module introduces us to the "verbs" of the subject, that is, the functions that allow us to move objects around in the plane or in space. In particular, in this module we explore the two flavors (elations and homologies) of one of the most useful of such functions: the perspective collineation.

We will start with a review of two exercises that we did earlier in the semester.

1. Figure 8.1 shows the start of a drawing of the letter T in two-point perspective; we give both vanishing points and one corner of the back face. Draw the rest of back face of the T. Assume the edges are solid and the faces are transparent, so we can see all 24 edges.

2. Figure 8.2 shows a vertical, rectangular fence panel receding into the distance. Draw another fence panel whose near corner is at the point P; the second panel should be the same size and shape (perspectively speaking) as the first panel.

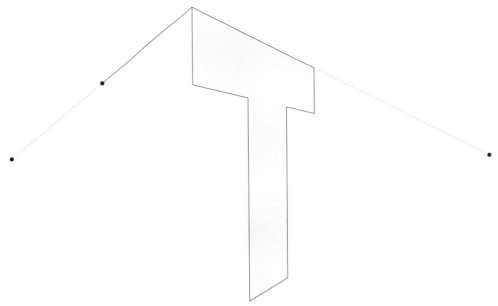

FIGURE 8.1: Finish drawing the transparent T in perspective.

FIGURE 8.2: Draw another fence panel, the same "size" and "shape," whose near corner is at the point *P*.

Let us connect these review exercises to Desargues's theorem.

3. In each picture above, create a triangle and its image; for example, you can create a triangle on the front face of the T by connecting the top near corner of the T to the two points at the farthest edge of the T. Then do the same for corresponding points on the back face.

4. Are your triangles in the T perspective from a point? If so, which point?

5. Are your triangles in the T perspective from a line? If so, which line?

6. Are your triangles in the fence perspective from a point? If so, which point?

7. Are your triangles in the fence perspective from a line? If so, which line?

It is time to learn some formal vocabulary.[1] Read the definitions below aloud to your group. If there are words your group does not understand, ask your professor about them.

> **Definition** A *perspective collineation* is a function $h : \omega \to \omega'$ between planes $\omega, \omega' \subset \mathbb{E}^3$ (possibly the same plane) that takes points to points and lines to lines, satisfying both of these conditions:
>
> - there exists a point $O \in \mathbb{E}^3$ (called the *center*) such that O, P, and $h(P)$ are collinear for every point $P \in \omega$, and
> - there exists a line $o \subset \omega \cdot \omega'$ (called the *axis*) such that o, ℓ, and $h(\ell)$ are coincident for every line $\ell \subset \omega$.
>
> If $h(O) \in o$, then we say h is an *elation*; otherwise h is a *homology*.
>
> We say the collineation *fixes* a point P or a line ℓ (or that P or ℓ is *fixed*) if $h(P) = P$ or $h(\ell) = \ell$.

Both the T and the fence show perspective images of translations; but they are different kinds of translations.

8. The T is translated from one plane to another (the front face to the back face). Does the T construction show an elation or a homology?

---

[1] We did not make this terminology up, although it could have been even stranger. Girard Desargues, in his work introducing the ideas of projective geometry, used botanical terminology for common words; the mathematician Julian Lowell Coolidge once commented that Desargues's "style and nomenclature are weird beyond imagining."

9. The fence is translated within the same plane. Does the fence construction show an elation or a homology?

Now let us consider other ways of moving things around in a plane. Figure 8.1.0 shows a drawing of a floor with equilateral triangular tiles. Six shaded triangles have been given labels and markings, so that we can try to match up the corresponding lines via a unique mapping.

10. Triangle $\mathcal{E}$ is the reflection of $\mathcal{T}$ in a mirror on the vertical wall. On Figure 8.1.0, shade in the image of the reflection of triangle $\mathcal{B}$ in that mirror.

11. Using the table on the next page, think about the various mappings that take us from triangle $\mathcal{T}$ to each of the other six triangles $\mathcal{A}, \mathcal{B}, \mathcal{C}, \mathcal{D}, \mathcal{E}$, and $\mathcal{F}$. For each of these mappings, determine whether the mapping shows the image of a translation, a rotation (if so, by how much), a reflection, a dilation, or a glide reflection.

12. Building on question 11, in the remainder of this table, for each of these mappings, determine whether the picture shows a perspective collineation, and if so,
    (a) where is the center O?
    (b) where is the axis o?
    (c) does the figure show a homology or an elation?
    Answer these questions first for the image map h' with $h'(\mathcal{T}') = \mathcal{X}'$; you might draw directly on Figure 8.1.0 to answer this part. Then answer the same questions for the corresponding real-world map h with $h(\mathcal{T}) = \mathcal{X}$.

Table for questions 11 and 12

Mapping	$\mathcal{T} \to \mathcal{A}$ or $\mathcal{T}' \to \mathcal{A}'$	$\mathcal{T} \to \mathcal{B}$ or $\mathcal{T}' \to \mathcal{B}'$	$\mathcal{T} \to \mathcal{C}$ or $\mathcal{T}' \to \mathcal{C}'$	$\mathcal{T} \to \mathcal{D}$ or $\mathcal{T}' \to \mathcal{D}'$	$\mathcal{T} \to \mathcal{E}$ or $\mathcal{T}' \to \mathcal{E}'$	$\mathcal{T} \to \mathcal{F}$ or $\mathcal{T}' \to \mathcal{F}'$
Type of mapping (for real-world map)						
Perspective collineation? (yes/no)						
Center (for image map)						
Axis (for image map)						
homology/elation						
Center (for real-world map)						
Axis (for real-world map)						
homology/elation						

As you can see from the examples above, perspective collineations often come up naturally when we think about moving things around in three-dimensional space. So the following two theorems, each of which gives a kind of "minimum amount of information," are very useful for understanding the mathematical side of this subject.

Three Points Theorem: *A perspective collineation is determined by three points (non-collinear) and their images.*

Center-Axis-Point Theorem: *A perspective collineation is determined by its center, its axis, and a point not on the axis together with its image.*

Let us consider how we might prove these theorems. Figure 8.3 shows three points and their images under a perspective collineation. Use a ruler and pencil to add lines and points to Figure 8.3, and as you do so, answer the questions below.

13. Assume we are given three points A, B, and C in $\mathbb{E}^3$, not collinear, and their images A', B', and C'. How do we determine the location of the center O?

14. How do we determine the location of the axis o?

15. Now choose any other point $P \in \mathbb{E}^3$, coplanar with ABC. Describe a method for determining the location of P'.

16. What is the image of the line AA'?

17. Let $Q = AB \cdot o$. Where is Q'?

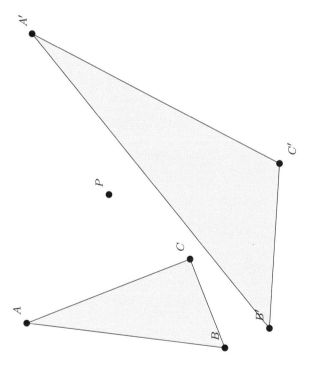

FIGURE 8.3: A perspective collineation sends points A, B, and C to their images A', B', and C'. Where is the center? Where is the axis? Where is P'?

18. Draw a line ℓ on the picture. Describe a method for determining the location of ℓ'.

Since P and ℓ are arbitrary, our argument shows that the perspective collineation is completely determined by A, B, C, and their images. This proves the three points theorem.

Now we turn to the center-axis-point theorem. Suppose we have an a center O, an axis o, a point X, and its image X', as in Figure 8.4.

19. Given a point Y coplanar with X and o, describe a method for determining the location of Y'.

20. Given a line k coplanar with X and o, determine the location of k'.

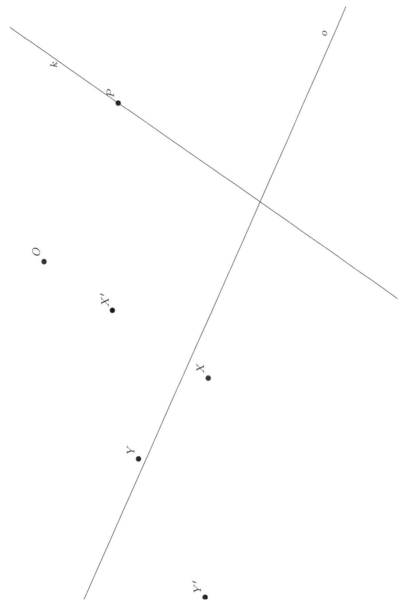

FIGURE 8.4: Given the center O, the axis o, and one pair of corresponding points X and X', describe a method for determining Y' and k'.

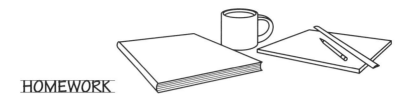

<u>HOMEWORK</u>

EXERCISES

Ⓔ 8.1.1. In what ways does the definition of a mesh map differ from that of a perspective collineation? When is a mesh map also a projective collineation?

ART/MATH ASSIGNMENT

◇ 8.1.1. Inspired by a question about meshes, George and Naije designed a logo with exactly four points and four lines. They drew a version of the logo on the vertical wall in Figure 8.5, with one of the four lines of the logo also vertical. Liking what they saw, they drew two other copies of the logo, exactly the same size, one translated and one rotated. When George snapped a photo of the wall, Naije noticed that the original logo was reflected off of a puddle on the horizontal ground.

Your Art/Math Project will be to complete the drawing of the photograph with the three other versions of the logo. You should turn in one "neat" copy of the drawing (with no construction lines); you should also turn in one or more drawings that show your constructions, clearly showing the center and axis that you use to construct the images.

(a) Naije and George translate the logo so that the point P moves to the point T. Where is the center of this perspective mapping in this photograph? Where is the axis of this mapping?

(b) Naije and George rotate the logo 180° so that the point P moves to the point R. Where is the center of this perspective mapping in the photograph? Where is the axis of this mapping?

(c) Naije sees a reflection of the logo mirrored in a puddle on the ground; the reflection of the point P is the point M. Where is the center of this perspective mapping in the photograph? Where is the axis of this mapping?

The constructions should be neat and clear enough that your steps are understandable.

In addition to the neat copy and the constructions above, for the reflection of the logo in the puddle, you should include a top view and a side view that show points on the logo, points in the reflection (in the puddle), and points in the images of the logo and its reflection. Use these top and side views to explain both the center and axis of the real-world map, and also the center and axis of the image map within the picture plane.

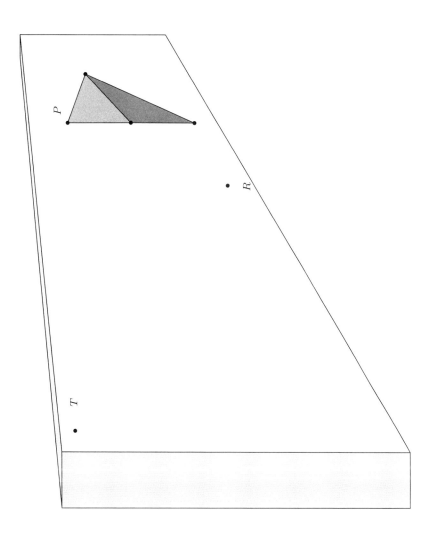

FIGURE 8.5: A logo with four lines (one of them vertical) and four points. Can we create *Rotated*, *Translated*, and *Mirror* images of this logo?

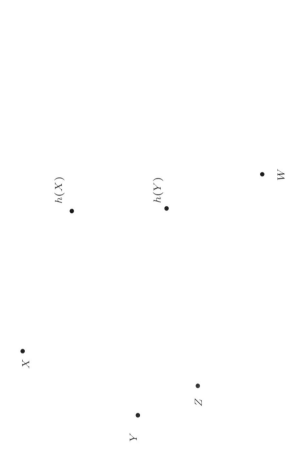

FIGURE 8.2.0: Four points in $\mathbb{E}^2$ and the images of two of them, under a period-2 homology. [For use with the REFLECTING ON HOMOLOGIES AND HARMONIC SETS module.]

8.2

Reflecting on Homologies and Harmonic Sets

Overview In the HOW PROJECTIVE GEOMETRY FUNCTIONS module, we saw that a translation from one plane to another gives us a form of perspective collineation, but a translation within the same plane is particularly special, because it gives us an *elation*.

In the same way, a reflection from one plane to another gives us a form of perspective collineation, but a reflection within the same plane is particularly special, because it gives us a *period-2 homology*.

Before we begin our exploration, we'll remind ourselves of old definitions and introduce new ones. You don't need to read these definitions carefully before you start the problems; we just collected all relevant definitions here in one place for ease of finding them as you need them.

Definition A *quadrangle* is a planar mesh containing exactly four points, no three of which are collinear, and the six lines determined by pairs of the points. We say that two lines of a quadrangle are *opposite* if they do not share any of the four points.

A set of four distinct collinear points A, B, C, D is called a *harmonic set*—which we write as $H(AC, BD)$—if there exists some quadrangle for which one pair of opposite lines are coincident at A, another pair of opposite lines are coincident at C, and the remaining two lines contain B and D respectively.

A *perspective collineation* is a function $h : \mathbb{E}^2 \to \mathbb{E}^2$ that takes points to points and lines to lines, satisfying both of these conditions:

- there exists a point $O \in \mathbb{E}^2$ (called the *center*) so that O, X, and $h(X)$ are collinear for every point $X \in \mathbb{E}^2$, and
- there exists a line $o \subset \mathbb{E}^2$ (called the *axis*) so that o, ℓ, and $h(\ell)$ are coincident for every line $\ell \subset \mathbb{E}^2$.

A *homology* is a perspective collineation for which the center does not lie on the axis. A *period-2 homology* is a homology h for which $h(h(X)) = X$ for every $X \in \mathbb{E}^2$.

Suppose h is a homology with center O and axis o. For every $X \in \mathbb{E}^2 \setminus \{O\}$, we denote by X_o the intersection of the lines o and OX. We say h is a *harmonic homology* if the points $X, X_o, h(X), O$ form a harmonic set: that is, if $H(Xh(X), X_oO)$.

1. For the points in Figure 8.7, we claim that $H(AC, BD)$. Draw a quadrangle that verifies this claim.

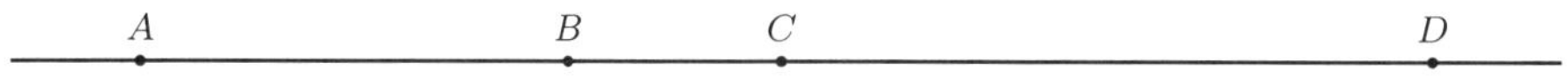

FIGURE 8.7: $H(AC, BD)$.

2. How do harmonic sets relate to perspective drawings?

3. We replicate Figure 8.7 below in Figure 8.8. Demonstrate that $H(BD, AC)$. (You may do so in either Figure 8.7 or Figure 8.8, as you prefer.)

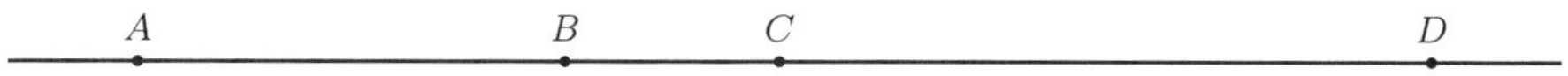

FIGURE 8.8: $H(AC, BD)$.

4. For Figures 8.7 and 8.8, is it true that $H(AB, CD)$?

In problems 5–10 below, we assume h is a period-2 homology, and we choose X and Y in $\mathbb{E}^2$ so that $X, Y, h(X), h(Y)$ are the four points of some quadrangle. As you answer these questions, you might wish to follow along with Figure 8.2.0.

5. One of the lines in the quadrangle mentioned above is the line XY. Name all six lines of the quadrangle and their corresponding images under h. Use the fact that h is period 2; for example, we can say $hh(XY) = XY$.

line	h(line)

6. Use your answers to the previous problems to give a method for constructing the location of the center O of the homology h. Locate O in Figure 8.2.0.

7. Use your answers to the previous problems to give a method for constructing the location of the axis o of the homology h. Locate o in Figure 8.2.0.

14. Locate a quadrangle that acts as the perspective image of a rectangle for which Y and $h(Y)$ are the primary vanishing points.

15. Determine the vanishing points of the diagonals of the quadrangle you named in the previous problem. How do you know these points are collinear with Y and $h(Y)$?

16. Therefore [**here and in the next problem comes the big ta-dah . . .**], the points $Y, Y_o, h(Y), O$ form a [_____].

17. We have just proved that every _____ homology is a _____ homology.

18. Using the same picture, adding no new lines and only the point $X_o = o \cdot OX$, locate a quadrangle that shows that $\mathsf{H}(Xh(X), OX_o)$. (This is tricky!)

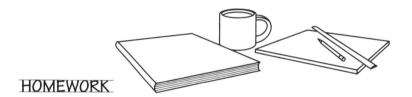

HOMEWORK

EXERCISES

Ⓔ 8.2.1. Explain why a 180-degree rotation is a period-2 homology.

In Exercise 3.2 of the IMAGE OF A LINE module, we defined a *perspectivity*. The following questions build on this notion. [Here is the definition again: Let ℓ and ℓ' be two distinct lines in $\mathbb{E}^2$. Let $O \in \mathbb{E}^2$ be a point not on either of these two lines. For any point $P \in \ell$, we will let P' be the projection via O onto the line ℓ'; that is, P' is the intersection of the lines ℓ' and OP. This projection is called a *perspectivity*.]

Ⓔ 8.2.2. In Figure 8.9, there is a perspectivity taking points $X, X_o, h(X)$, and O to Y, Y_o, $h(Y)$, and O respectively. From what point does the projection take place?

Ⓔ 8.2.3. Argue using a similar diagram that the following statement is always true: Let $ABCD$ be four points on line ℓ such that $\mathsf{H}(AC, BD)$. Let ℓ' be a line distinct from ℓ such that $D = D' = \ell \cdot \ell'$. If A', B', and C' on ℓ' are the images of points A, B, and C through a perspectivity with center O, then $\mathsf{H}(A'C', B'D')$.

ⓔ 8.2.4. Use Figure 8.10 to prove that a harmonic set is a projective invariant, by drawing the line $\ell'' = B'D$ and using the lemma.

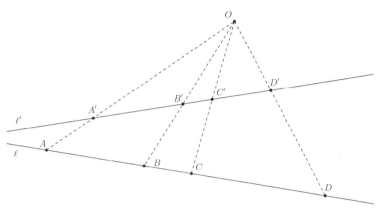

FIGURE 8.10: Figure for Exercise ⓔ 8.2.4

PROOF/COUNTEREXAMPLE

ⓟ 8.2.1. Prove the converse of statement 17.

FIGURE 8.3.0: The perspective image of a translation in a plane. [For use with the ELATIONS (OR, HOW TO BE TRANSPORTED IN A MATH CLASS) module.]

7. Locate the point $e(C)$.
8. By repeating the ideas from the last few steps, sketch the image of the translated triangle $e(ABC) = e(A)e(B)e(C)$. Check your solution by making sure the triangles are perspective from both the point O and the line of the horizon.
9. Draw the image of the translation of the translation: that is, draw $e \circ e(ABC)$.
10. Continue this process. Draw several more images of translated triangles receding into the distance.
11. Now draw the images of several translations of the point X.
12. Explain why Figure 8.3.0 shows an elation.

13. How does the axis of this elation relate to the axis of the original translation in the ground plane?

14. How does the center of this elation relate to the center of the original translation in the ground plane?

15. Fill in the blank to complete the following theorem:

> **Elation Theorem.** *An elation is completely determined by its axis, by its center, and by _____ pair(s) of corresponding points.*

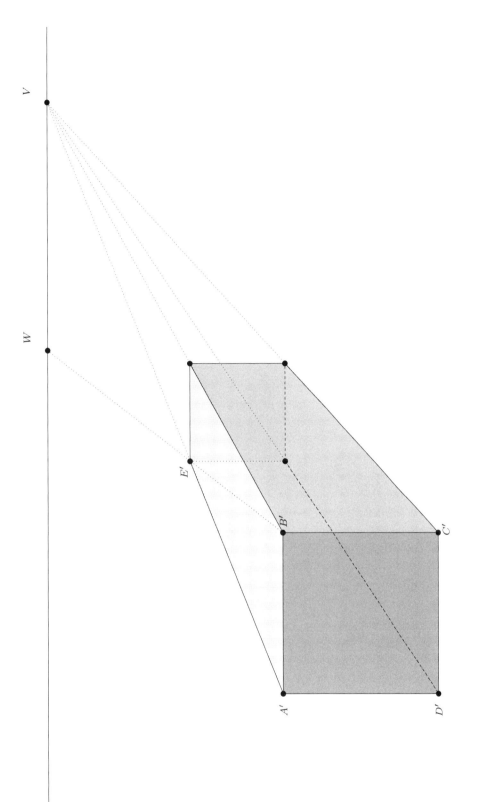

FIGURE 9.0: What shape is this box? Is it a cube? Or is it twisted? [For use with the DYNAMIC CUBES AND VIEWING DISTANCE module.]

9

Dynamic Cubes and Viewing Distance

Overview In the first part of this module, we will get a bit more practice using GeoGebra.

But even more importantly, the second part of the module allows us to discover a mathematical method for determining viewing distance. When you taped windows, you learned that you could figure out where the Art Director stood; you probably did this by trial-and-error, moving yourself around until lines you saw on the window matched up with lines you saw outside. In this module, we'll learn one way to determine the correct viewing location using simple geometry instead of trial-and-error.

The steps 1–14 below will help us to draw a picture in GEOGEBRA that looks like Figure 9.0.

1. Open up GEOGEBRA and turn off the axes (and, optionally, labeling). If you wish to have the label of your points match what follows, keep the labeling option on. When you create a new point, "control-click" on the point and select "rename."

2. In GEOGEBRA, create two points: one will be the main vanishing point (as with Figure 9.0, here, we will call it V); the other will be the vanishing point (W) of the diagonal on the top of the box.

3. In GEOGEBRA, connect V and W with a line. For reasons you will describe in the exercises, we will call this line h for "horizon," even though your version of h will probably not be horizontal (and that's okay). Even so, in what follows, we will use the convention

that "side-to-side" means directions parallel to h and "up/down" means directions perpendicular to h.

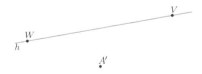

4. Now we will begin drawing our box. Add another point (A') somewhere in the picture. Then, using the drop-down menus to get to the "perpendicular" and "parallel" line tools, we will draw two lines through A': one perpendicular to $h = VW$ and one parallel to h. These lines will give us two of the front edges of our box.

5. Next, to find two equally spaced points, we'll use the circle tool to create a circle centered at A' and passing through another point B'

Definition For a perspective picture, the *viewing target T* is the point in the picture plane ω that is closest to the center O. That is, OT is perpendicular to ω. The *viewing distance* $|OT|$ is the distance from the center to the picture plane.

From here forward, let us assume that the box you drew is actually the perspective image of a cube. The following two questions ask for the side view and top view that correspond to the cube you drew (or likewise to Figure 9.0).

15. As in the questions above, we will use the convention that "left/right" means directions parallel to h and "up/down" means directions perpendicular to h.
    In the space below, draw a side view of the cube that includes the points A, B, C, D, E, V, W, and O. As you add these points, answer the following questions.
    (a) Which (if any) of these points must be the same in this side view?
    (b) How does the line h ($= VW$) relate to the cube in this side view?
    (c) How does the line OV relate to the cube in this side view?
    (d) How does the line OW relate to the cube in this side view?
    [Note that A is a point on the "real-world" cube; A' is the image that you drew in the steps above. This question asks you to include A but not A' in your side view.]

16. In the space below, draw a top view of the cube. Include the points A, B, C, D, E, V, W, and O, and answer the following questions.
    (a) Which (if any) of these points are the same in this top view?
    (b) How does the line h ($= VW$) relate to the cube in this top view?
    (c) How does the line OV relate to the cube in this top view?
    (d) How does the line OW relate to the cube in this top view?

17. Where is the viewing target T in the above side and top views?

18. Using similar triangles in one of your side and top views above, explain why the viewing distance $|OT|$ is equal to the distance between the vanishing points V and W: that is $|OT| = |VW|$.

19. Consider Figure 9.0. Look at this picture with one eye leave one inch of space for response in front of the point V, with your eye very close (a distance $|VW|$ from the point V). What do you see? Is it a dumpster? a cube?

20. Add the following words (*artist, canvas, perspective, viewer*) into the sentence:

    Accurate _____ viewing isn't just about how the _____ created the drawing or painting, but also about where the _____ stands when looking at the _____.

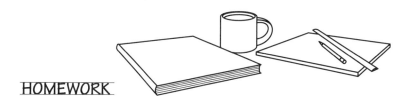

HOMEWORK

EXERCISES

ⓔ 9.1. Explain why we might reasonably assume, for a real-world cube and a vertical picture plane, that the line VW would be horizontal.

ⓔ 9.2. Dürer's *Saint Jerome in His Study* (Figure 9.1) is widely considered to be his "show-off" piece that demonstrated his facility with linear perspective. In the 1930s, *Saint Jerome* came under heavy criticism from an influential curator who claimed the perspective in this etching was riddled with errors. For example, William M. Ivins Jr. wrote, "The top of the saint's table is of the oddest trapezoidal shape—certainly it is not rectangular." Assuming that Ivins was wrong and that Dürer correctly drew the table as the image of a square (see, for example, the defense of Dürer's work in [15]), determine the viewing target and viewing distance for this etching.

ⓔ 9.3. In 2009, Professor Andrius Tamulis of Governors State University snapped a photograph of an unusual painting on a truck (see Figure 9.2). This type of painting is called *trompe l'oeil* (pronounced "tromp loy"), French for "fool the eye." A trompe l'oeil painting is a perspective painting that attempts to trick the viewer into thinking the flat surface of the painting is not there, and that the viewer is looking at real depth. Although the photograph of the truck is in two-point perspective, the painting on the back of the truck is in one-point perspective. What can we say about the viewing target and viewing distance?

   (a) Is the intended viewer likely to be another motorist (say, the driver of another car), a dockworker standing on a loading platform, or someone else? How can you tell?

   (b) According to the YRC yrc.com/trailer-dimensions, the bottom of the doors of the truck are 50" from the ground, the truck stands 13'6" high and is 45' long. Draw a side view and/or top view that shows how we could calculate viewing distance from this information.

FIGURE 9.1: *St. Jerome in His Study*, Albrecht Dürer, 1514, etching, 10″×7″ [18].
Courtesy of SLUB Dresden / Deutsche Fotothek / Rudolph Kramer

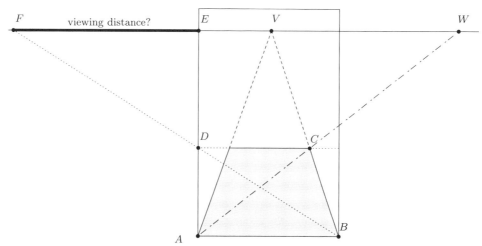

FIGURE 9.4: Alberti's construction of the viewing distance of a square in one-point perspective.
Courtesy of Professor Andrius Tamulis

diagonal of the square AC out to the horizon to get W (as we did), he draws a line through C parallel to the horizon and extends this to the edge of the picture (point D). He draws the line BD, which intersects the horizon at point F. He claims that the distance $|EF|$ from F to the edge of the canvas is the viewing distance [42].

If it is true that $|EF| = |VW|$, give a proof of this equality. If not, provide a counterexample with an explanation of why your example contradicts Alberti's claim.

D_2 V_L D_1 V_R

FIGURE 10.0: Is this shape the image of a rectangle? Of a square? We'll find out! [For use with the DRAWING BOXES AND CUBES IN TWO-POINT PERSPECTIVE module.]

10

Drawing Boxes and Cubes in Two-Point Perspective

Overview In the *Dynamics Cube* module, we learned how to determine viewing distance and how to draw cubes in one-point perspective. This module bumps our ability up one point: we'll look at how to determine viewing distances for boxes in two-point perspective. When we get good at that, we'll draw the image of a two-point perspective cube.

WARM-UP EXERCISE. Figure 10.0 shows the image of a rectangle (possibly even a square?) in two-point perspective. The goal of this module will be to figure out how to draw the perspective image of a cube for which the shaded region is the top face, and to understand how to determine viewing distance for a two-point perspective picture.

1. Use a straightedge to verify that the points V_L and V_R are vanishing points of the edges of the shape in Figure 10.0, assuming that the edges of the original object are mutually parallel.
2. Use a straightedge to determine the significance of the points D_1 and D_2. What role do these points play in the image?

If the object in Figure 10.0 is the image of a rectangle, then our two lines of sight to the two vanishing points V_L and V_R ought to meet at our eye at a $90°$ angle.

3. Lay another regular piece of paper down so that the edges lie exactly on V_L and V_R. Put a dot on the drawing where the corner of the paper is; if you do this correctly, then the line between the dot and V_L should be perpendicular to the line between the dot and V_R.
4. You can repeat the step above with the paper in another orientation, with the edges of the top paper still lying exactly on the vanishing points, but the corner in another location. Add a second dot with this new orientation.
5. Repeat this process again and again. You will probably see a curve emerging from your dots.
6. What is the name of the curve that passes through all your dots?

Imagine tilting that curve up and out of the plane so that the curve is perpendicular to the picture plane. We will call this curve a *viewing circle*; our construction tells us that if we want to see the pictured object as the image of a rectangle, we need to look at it with our eye level with the horizon, located somewhere on the semi-circle that connects the two vanishing points.

7. Draw a top view showing the viewer, the picture plane, the vanishing points V_L and V_R, and the viewing circle.

8. If the object is not merely a rectangle, but is actually a square, what can you say about the angle formed by the diagonals of the square (in the actual square, not in the image)?

9. On Figure 10.0 construct a second viewing circle, similar to the first, but this time using the points D_1 and D_2.
10. Add the lifted version of this circle to your top view. What does this top view tell us about the location of the viewer?

11. We know that to see the object in Figure 10.0 as a square, we view the image from somewhere directly across from the horizon. What do these two viewing circles tell us about where to stand? Make two notations on the picture:
    (a) the *viewing target T*, which is the point on the picture plane nearest to the viewer, and
    (b) the *viewing distance*, which we can indicate by a line segment in the picture plane that has the same distance as the distance between T and the viewer.

Now that we've had a good warm-up, we'll move into the big ideas of this module by drawing the image of some rectangular box in two-point perspective. When we get good at that, we'll figure out the slightly harder question of how to draw the image of a cube.

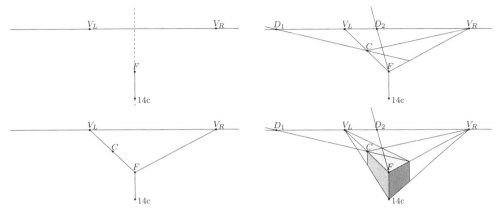

FIGURE 10.1: Illustrations that will help with steps 12–20 below.

Warning! *Part of the construction on the next page is actually wrong and/or impossible! But the mistake below is very easy to make, and part of the goal of this module is that you discover what that mistake is as you go along. Fixing that mistake will lead us to a fundamental (and very cool) property that relates geometry to numbers. So keep your wits about you.*

---

12. Fire up GEOGEBRA; get rid of axes and (optionally) turn off labeling.
13. Create a line through two points. (If you want, you can adjust the points until the line is horizontal according to the algebra window, but this step is not necessary.) Hide the original points. Then add to this line two vanishing points, which we'll call V_L and V_R.
14. Now we will go through steps we learned when drawing a one-point perspective cube; first we'll draw the near vertical edges of our box:
    (a) Add a point not on the horizon (we'll call that F for "front"; it will be the top front corner of your box).
    (b) Add a line through that point and perpendicular to the horizon. (Do *not* just eyeball this; use the "perpendicular" tool!)
    (c) Choose another point on that "vertical" line (vertical is in quotes because your horizon line might not be horizontal, and that's okay).
    (d) Hide the vertical line.
    (e) Add a line segment between the last two points you created.
    The front vertical edge of your box is now done.
15. Add lines from the top and bottom of this edge to your existing vanishing points V_L and V_R.
16. Choose a point on one of the lines that connect to F to V_L; that new point will be a third corner of your box (I'll call that C for "corner").
17. Now we'll add two points D_1 and D_2 on the horizon that are the vanishing points of the diagonal of our box.
18. Connect D_1 to F and connect D_2 to C.

19. From here, finish the box by adding intersection points and by drawing lines (or line segments) that go either to a vanishing point or are perpendicular to the horizon.
20. When your box is done, hide all construction lines; leave visible the horizon and the segments or polygons that make up your box.

The exercise above leads us to two interesting questions:

- How do we draw a cube (or really, how do we draw a box of any given dimensions)?
- What is the relationship among V_L, V_R, D_1, and D_2?

The former question, it turns out, is a geometry question, and the latter is a projective geometry question. So we'll handle these questions separately.

To help us answer one or the other of these questions, there's one last thing for you to do with your GeoGebra worksheet.

Once you fix the mistakes above and create your box with appropriate vanishing points V_L, V_R, D_1, and D_2, unhide the original points you used to make your horizon line and then use the algebra bar on the left side to try to get the horizon as horizontal as possible (that is, make sure the y-coordinates of those vanishing points are nearly identical).

21. Write down the x-values of the coordinates of each of those points:
    (a) $V_L =$
    (b) $V_R =$
    (c) $D_1 =$
    (d) $D_2 =$
22. Determine the following four distances:
    (a) $V_L - D_1 =$
    (b) $V_R - D_2 =$
    (c) $D_1 - V_R =$
    (d) $D_2 - V_L =$
    Don't lose those numbers. We'll use them later!
23. In your GeoGebra file, …
    (a) …use the semicircle tool to create a semicircle with endpoints at D_1 and D_2 and another with endpoints at V_L and V_R.
    (b) Locate the intersection of these semicircles; call it O.
    (c) Use the angle tool in GeoGebra to measure these angles: $\angle D_2 O V_L$, $\angle V_L O D_1$, and $\angle D_1 O V_R$. What do you notice?

Drawing a Cube

On the previous pages, you drew the image of some rectangular box. Is this box actually a cube? As before, the answer depends on where you stand … but in this case, it also depends on the length of the "vertical" segments.

So our next goal is to draw a cube—not just some random box—starting with only a few clues. To help you in this endeavor, we include construction steps in Figures A–J. Along the way, we will locate the correct viewing target T and viewing distance.

24. Your task is to first understand and then explain the construction steps at each stage. Write your description of what is happening next to each figure, as appropriate.

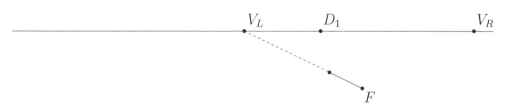

FIGURE A: Beginning stage.

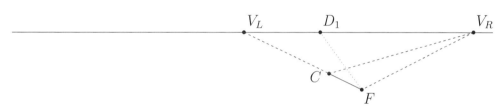

FIGURE B: Explain the dotted lines.

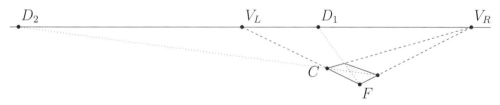

FIGURE C: How did we construct the top of the cube?

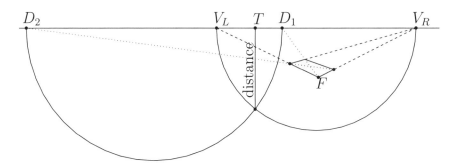

FIGURE D: Explain how we get the viewing target *T* and the line segment representing viewing distance.

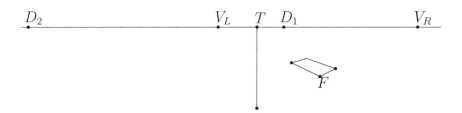

FIGURE E: Figure D with some construction lines and circles erased, so you can add new lines after you look at the next few figures. See also the larger figure from your "warm-up exercise."

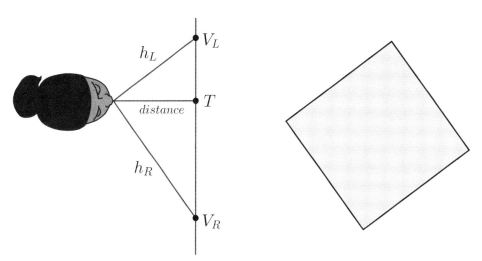

FIGURE F: Locate $V_L T$, $V_R T$, and *distance* in Figure E and in Figure 10.0 of the warm-up exercise. Then draw line segments on each figure showing the lengths h_L and h_R. (We use the symbol '*h*' for 'hypotenuse'.)

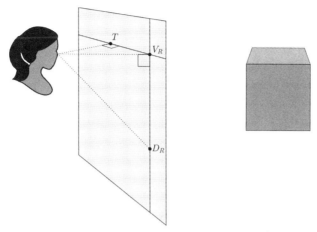

FIGURE G: Here, we look at the cube so that one of the sides is parallel to our view
(which means the picture plane is angled from this view).
The point D_R is the vanishing point for which line in the cube?
Where can we draw the hypotenuse h_R (from Figure F) in *this* picture?
(*Hint*: Because the faces of the cubes are all square, h_R appears in two places in this view.)

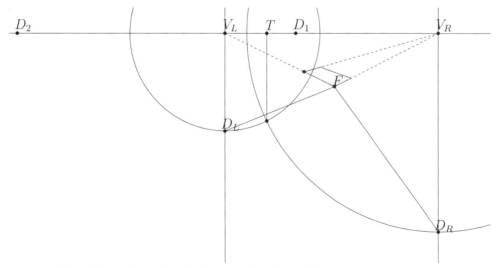

FIGURE H: How did we determine the locations for D_L and D_R?

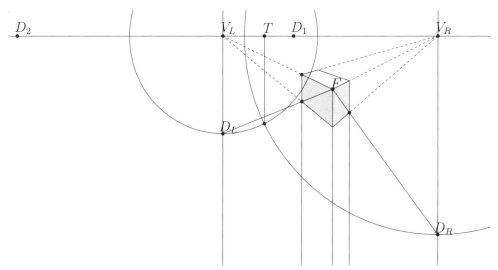

FIGURE I: How do we construct the two vertical faces of the cube?

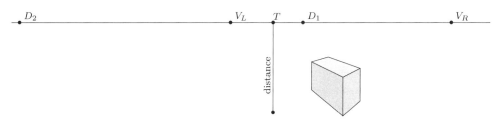

FIGURE J: The image of a cube, with viewing target *T* and viewing distance indicated.

25. Finally, use the construction above to draw the two-point perspective images of two cubes in Figure 10.0. One of the cubes should be the one whose top face already appears in the figure, and another can be at a location of your own choosing. (Nearer the horizon will look "better," and further from the horizon will look more distorted, but anywhere you draw the second cube, you should use the vanishing points V_L, V_R, D_1 and/or D_2, and D_L and/or D_R.)

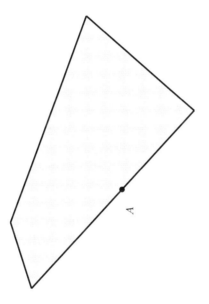

FIGURE 10.2: A drawing of a square in two-point perspective, for homework exercise 10.1.

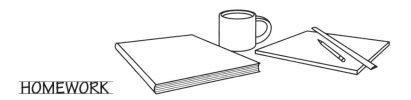

HOMEWORK

EXERCISES

ⓔ 10.1. In Figure 10.2, you see the image of a square in two-point perspective with a point A along one edge of the square. Use ruler-and-compass methods to draw a figure ABCD that is the image of another square, inscribed in the original, as in Figure 10.3. (*Hint*: You'll need to determine the vanishing line and vanshing points as well as viewing distance. The line AC should pass through the perspective center of the original square; the lines AC and BD are images of lines that are perpendicular in the real world.)

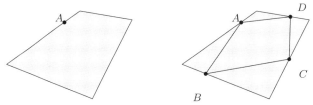

FIGURE 10.3: Draw the image of an inscribed square with one corner at the point A.

ⓔ 10.2. Given the two-point perspective image of a square in a horizontal plane, describe the construction that allows us to draw an octahedron by attaching the image of two squares in vertical planes. (See Figure 10.4.)

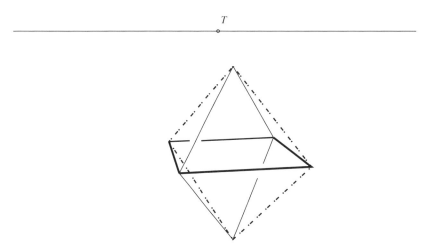

FIGURE 10.4: The perspective image of an octahedron composed of three squares: one horizontal and two vertical.

ART ASSIGNMENT

⚠ 10.1. Write a word that is at least four letters long in two-point perspective. The letters should be a constant width, the spaces between them should be a (smaller) constant width, and the depth of the letters should be consistent (see Figure 10.5). The width of the lines in the letters should appear to be consistent (the bar in a "T" is the same width as the bars in an "H", for example). Give the word a surrounding context (is it sitting on a table? mounted on the back wall of a room? in a vast plane with buildings on the horizon?)

Draw lots of lines; draw neatly; carefully and completely erase construction lines that you no longer need. You may be tempted to embellish your figure with decorations. You should do so: in fact, the more perspectively correct embellishments there are, the better the picture will look. But you must use a straightedge for *every* line you draw, and you should not draw any curves.

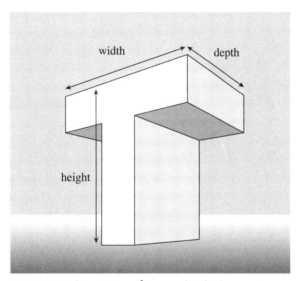

FIGURE 10.5: A prototype for your Art Assignment.

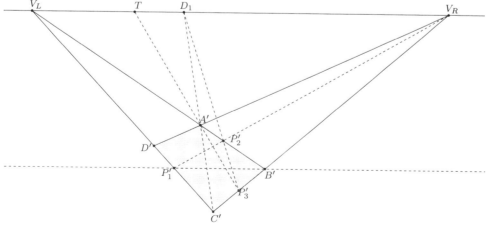

FIGURE 10.6: A construction for the Proof/Counterexample problem

PROOF/COUNTEREXAMPLE

[P] 10.1. Use side views and top views, as appropriate, to explain whether the construction in Figure 10.6 gives us the viewing target T, assuming $A'B'C'D'$ is the perspective image of a square $ABCD$. [We thank Robert Connelly of Cornell University for sharing this question with us.] The construction contains these four steps.

(a) Draw a line through B' parallel to the horizon $V_L V_R$; the intersection of this line with $C'D'$ determines the point P_1'.

(b) Locate the point $P_2' = (P_1' V_R) \cdot (A'B')$.

(c) Locate the point $P_3' = (D_1 P_2') \cdot (B'C')$.

(d) Let $T = (A'P_3') \cdot (V_L V_R)$.

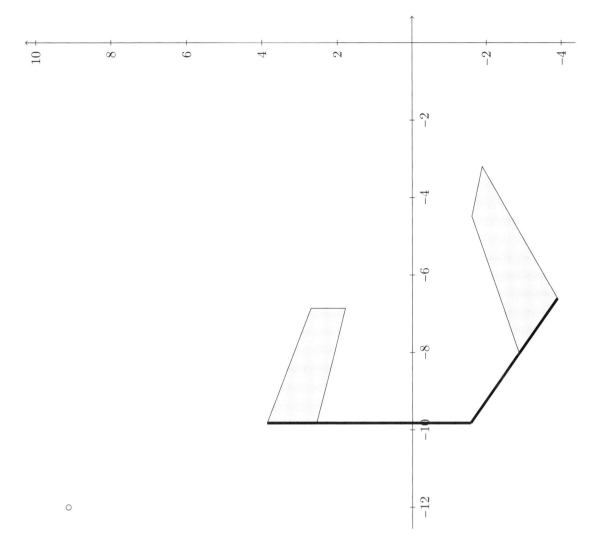

FIGURE 11.1.0: A flag and its shadow, each ready to be divided into three panels. [For use with the DISCOVERING THE CROSS RATIO module.]

11

Perspective by the Numbers

11.1 Discovering the Cross Ratio

Overview When we make perspective images of a pair of identical objects that are receding into the distance, the sizes of the images aren't identical. Is there anything at all we can measure that always stays the same after a perspective mapping?

It turns out, the answer is "yes," and in this module, we'll learn how to calculate and make use of the most famous and arguably the most powerful of these numerical invariants, the *cross ratio*.

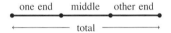

1. Given four equally spaced, collinear points (as above), what is the exact value of the following expression?

$$\frac{\text{(one end)}}{\text{(middle)}} \frac{\text{(other end)}}{\text{(total)}} = \underline{\hspace{2cm}}$$

2. The photograph in Figure 11.1 shows that images of equal-sized panels aren't themselves equally sized. Measure these lengths on that figure, and then compute the value of the following expression:

$$\frac{\text{(one end)}}{\text{(middle)}} \frac{\text{(other end)}}{\text{(total)}} = \frac{(\qquad)}{(\qquad)} \cdot \frac{(\qquad)}{(\qquad)} = \underline{\hspace{2cm}}.$$

How does your answer compare to the value you got in question 1?

FIGURE 11.1: The image of three equal panels, from a parking lot viewing target.

Definition Given any two ordinary points $A, B \in \mathbb{E}^3$, we denote the distance between these points by the symbol $\|AB\|$. We will use the symbol $|AB|$ to denote the *directed distance* between these points. That is, we assign (arbitrarily) a direction to the line containing the points; we say $|AB| = \|AB\| > 0$ if B comes after A, and $|AB| = -\|AB\| < 0$ if B comes before A. Note that $|AB|/|BA| = -1$ always.

Given four ordinary collinear points A, B, C, D, we define the *cross ratio* (AC, BD) by

$$(AC, BD) = \frac{|AB|}{|BC|} \cdot \frac{|CD|}{|DA|}.$$

3. What happens to the value of (AC, BD) if we change the direction in which we measure distance along the line?

4. The points E, F, G, and H lie on the x-axis as shown below. Compute (EF, GH).

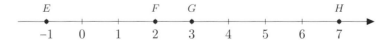

5. For the same diagram, compute (EG, FH).

6. The notation for the cross ratio can be confusing to remember; some people think of it as measuring segments that "cross over" the comma, as shown here.

Which of these (if any) is/are the same as (AC, BD)?

(a) (AB, CD) (b) (AC, DB) (c) (CD, AB) (d) (CA, BD) (e) (BD, AC)

Questions 1 and 2 seem to hint (and indeed eventually we will show that this is true) that the cross ratio of any image of four equally spaced, collinear points is always $-\frac{1}{3}$. For example, even without vanishing points or horizon lines, we can verify that the spacing of lines in the photograph of the track in Figure 11.2 is correct:

$$\frac{2.90}{0.499} \cdot \frac{1.118}{-(0.290 + 0.499 + 1.118)} = -\frac{1}{3}.$$

FIGURE 11.2: Markings on a running track.

7. One of the two sketches below shows a correct perspective image of gridlines on a football field; the other does not. Draw an extra line on each sketch that crosses the grid lines; measure and use the cross ratio to determine which sketch is correct.

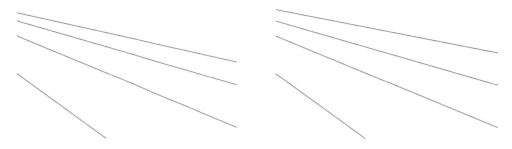

8. Suppose, as in question 4 above, we fix E at $x = -1$, F at $x = 2$, and G at $x = 3$, but we allow a point X on the x-axis to move closer and closer to the point at infinity. What happens to the ratio (EF, GX)? Does the answer depend on which direction (say, left or right) that we move X; if so, how?

9. More generally, suppose we fix A, B, and C on the x-axis, and again we move X closer and closer to the point at infinity. What happens to the ratio

$$(AB, CX) = \frac{|AC|}{|CB|} \cdot \frac{|BX|}{|XA|}?$$

Does the answer depend on which direction (say, left or right) that we move X; if so, how?

10. *Complete the definitions below, based on your answers to the above questions:*
    If ℓ is a line in $\mathbb{E}^3$ containing ordinary points P and Q and an ideal point $V_{[\![\ell]\!]}$, we will define
    $$\frac{|PV_{[\![\ell]\!]}|}{|QV_{[\![\ell]\!]}|} = \qquad \text{and} \quad \frac{|PV_{[\![\ell]\!]}|}{|V_{[\![\ell]\!]}Q|} = \qquad .$$

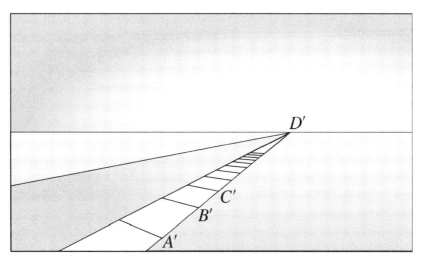

FIGURE 11.3: A perspective image of a sidewalk receding into the distance, for questions 11 and 12.

11. Consider Figure 11.3, which shows the perspective images A', B', and C' of three real-world points A, B, and C, and which also shows the vanishing point D' of the edges of the sidewalk. If the sidewalk lines are evenly spaced, what is the cross ratio (AC, BD)? (Use knowledge of the real world, not measurements from the picture).

12. We have said above that the cross ratio is a projective invariant: that is, if A, B, C, D are collinear points and A', B', C', D' are their projective images, then $(AC, BD) = (A'C', B'D')$. Verify this invariance in Figure 11.3 by measuring and comparing the resulting cross ratio to your answer for question 11.

13. Pull this all together comparing Figures 11.4 and 11.5, which use the artist's method and the cross ratio method for locating vanishing points. In Figure 11.4, the line l in the photo of the tile floor has a vanishing point, which an artist would typically locate by extending lines whose preimages are parallel to the preimage of l. The intersection point is the vanishing point.

But without the aid of other lines that meet at the vanishing point of l, most artists would be helpless in locating that point!

(a) In Figure 11.5, use a ruler, a calculator, and the cross ratio to locate the vanishing point of l.

(b) Compare Figure 11.5 with Figure 11.4 showing the tile floor and l, so that the two figures align exactly. Measure, or possibly hold them up to a bright light so you can see the figures superimposed. How close did you come to locating the vanishing point?

The cross ratio, drawing rectangles, and dividing fences

How does the cross ratio relate to material we learned earlier in this course?

You might recall from drawing a box in two-point perspective that we can't just place the four vanishing points of a rectangle wherever we want; there are restrictions. Not surprisingly, these restrictions mean that the cross ratio of a set of vanishing points is a special number! But what number is this? Let's figure it out.

14. In the plan view in Figure 11.6, we assume that the rectangle has one edge parallel to the picture plane. Using a straightedge, locate the vanishing points V_a, V_c, V_b, and V_d of the indicated lines. Which one of these vanishing points is "at infinity"?

15. Use geometric rules (not measuring) to determine the cross ratio $(V_a V_c, V_b V_d)$.[1] Explain why this cross ratio does not depend on the dimensions of this rectangle—in particular, it does not depend on the slope of the lines c and d.

The special condition that allows a set of points to form the vanishing points for the image of a rectangle (and that constrains the cross ratio to be a special number) motivates the following definition.

> **Definition** When collinear points A, B, C, D satisfy $(AC, BD) = -1$, we say these points form a *harmonic set*; equivalently, we might write $H(AC, BD)$.

16. In Figure 11.7, verify that $(AC, BD) = -1$, and then demonstrate that these four points are the vanishing points for some perspective image of a rectangle. That is, use a straightedge to draw a rectangle in two-point perspective that has these four vanishing points for the edges and diagonals. If you assume moreover that the object that you drew is the perspective image of a square, then you should be able to locate the viewing target and draw the viewing distance.

---

[1]Note, by the way, that the cross ratio notation expressed this way keeps the main vanishing points together and the diagonal vanishing points together, which gives a bit of insight into why the notation, as cumbersome as it is, persists.

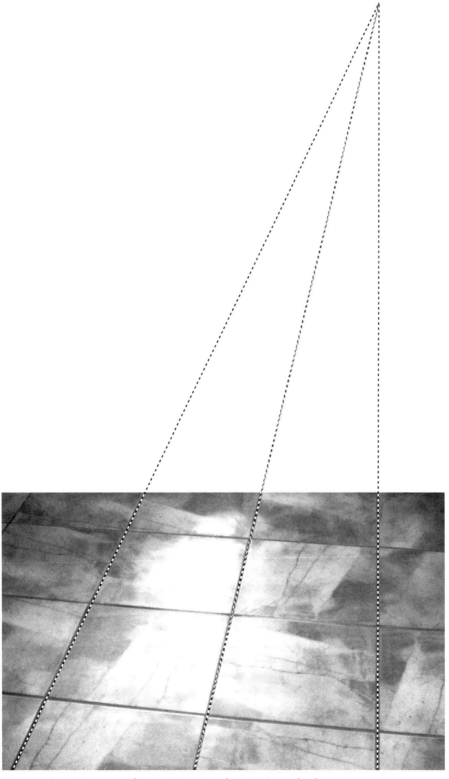

FIGURE 11.4: Locating a vanishing point using the artist's method.

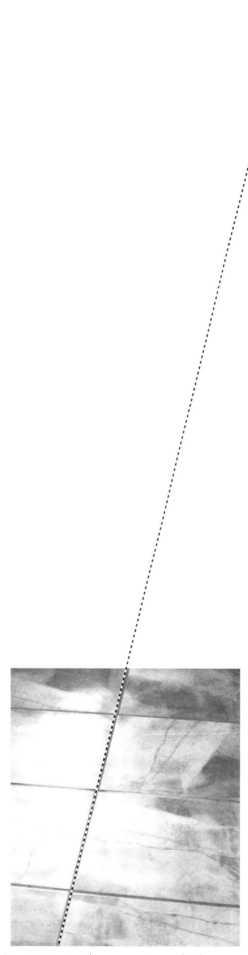

FIGURE 11.5: Locating a vanishing point using the cross ratio method.

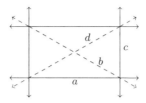

FIGURE 11.6: Diagram for questions 14–15.

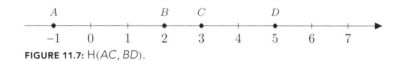

FIGURE 11.7: H(AC, BD).

17. Return to the photograph of the tile floor in Figure 11.5. Because we know that the three intersection points of the tiles with ℓ, together with the vanishing point, form a harmonic set, we could reinterpret these points as the vanishing points of the perspective image of a rectangle. Turn the paper sideways and imagine that the line ℓ is a horizon; use a straightedge and the method of constructing a perspective image of a rectangle to locate the vanishing point of the line ℓ.

Now we return to a question from an earlier module about dividing a rectangular fence panel into fractions.

18. In Figure 11.8, we assume that the leftmost panel has $1/n$ the width of the entire panel. If line a has slope 0 and line d has slope 1, then what is the slope of the line f? (*Hint*: You might begin by assuming the rectangle is one horizontal unit by one vertical unit. These units may or may not be equal: the rectangle might be one mile wide and one inch high.)

19. Locate the vanishing point V_f of the line f in Figure 11.8 and determine the cross ratio $(V_a V_c, V_b V_f)$.

20. Complete the statement of the following theorem: "Given a perspective image of a rectangle with edge vanishing points V and V' and diagonal vanishing points W and W', the image of a rectangle that has width $1/n$ of the original rectangle has edge vanishing points _____ and _____ and diagonal vanishing points points W and W_n. The cross ratio $(VV', WW_n) =$ _____ or _____, depending on the order of W and W_n.

The proof of the statement in question 20 uses our specific instance (above), with the magic-wand fact that the cross ratio is a projective invariant. We'll learn how to prove the

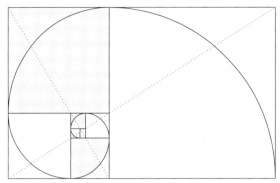

FIGURE 11.10: In the Golden Rectangle, as the diagonal lines demonstrate, each non-square rectangle is proportionate to the original rectangle. People often "decorate" the rectangle by adding quarter circles within the squares to create an approximation of the *golden spiral*.

compass, create a correct two-point perspective image of a golden rectangle, beginning with the large square and adding on additional rectangles which you can then subdivide.

PROOF/COUNTEREXAMPLE

P 11.1.1. An artist sketches the images of a pair of nested rectangular boxes in three-point perspective, as in Figure 11.11. The two boxes are oriented in the same direction and share the same near corner A. As a guide to drawing, the

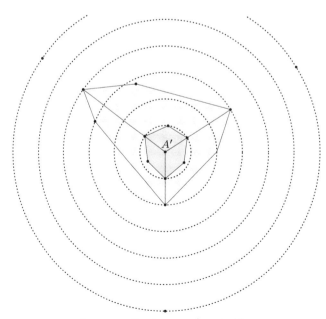

FIGURE 11.11: A perspective image of nested boxes.

artist draws evenly spaced concentric circles; these circles are centered at the image A' of the near corner.

If the volume of the smaller of the two boxes is 8 cubic units, determine the volume of the larger box, and explain why your answer is correct.[2]

---

[2] This question was inspired by Alec Regulinski, F&M Class of 2013, who after college went to graduate school in computer engineering at Cornell University. He taught a class on computer vision there, and included a question similar to this one on the final exam he gave his students.

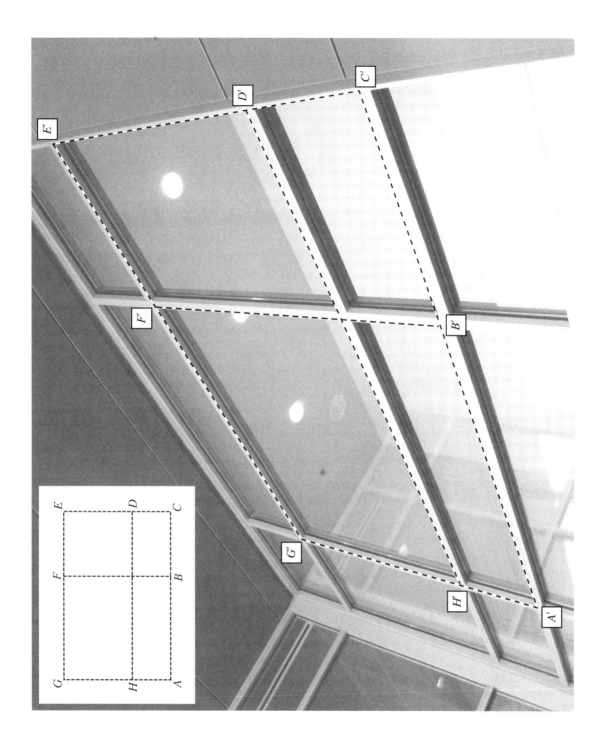

FIGURE 11.2.0: Photograph of a store window. [For use with the EVES'S THEOREM module.]

11.2

Eves's Theorem

Overview *The value of an h-expression is invariant under any projective transformation.*

Howard W. Eves, 1913–2000

The goal of this module is to give some insight into why G. C. Shephard [48] wrote, "We feel that Eves's theorem has never been given the recognition it deserves and should be regarded as one of the fundamental results of projective geometry."

---

1. For the diagram below (which also appears as an inset in Figure 11.2.0), give the exact value of the "circular product"

$$\frac{||AB||}{||BC||} \cdot \frac{||CD||}{||DE||} \cdot \frac{||EF||}{||FG||} \cdot \frac{||GH||}{||HA||}.$$

You should not need to measure; use geometry and logic.

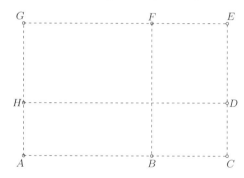

2. In the photograph in Figure 11.2.0, measure the dashed lines to determine the following quantities:

   (a) $||A'B'|| = $ _____ mm; $||B'C'|| = $ _____ mm; $\frac{||A'B'||}{||B'C'||} = $ _____;

   (b) $||C'D'|| = $ _____ mm; $||D'E'|| = $ _____ mm; $\frac{||C'D'||}{||D'E'||} = $ _____;

   (c) $||E'F'|| = $ _____ mm; $||F'G'|| = $ _____ mm; $\frac{||E'F'||}{||F'G'||} = $ _____;

   (d) $||G'H'|| = $ _____ mm; $||H'A'|| = $ _____ mm; $\frac{||G'H'||}{||H'A'||} = $ _____.

3. Compute the circular product:

$$\frac{||A'B'||}{||B'C'||} \cdot \frac{||C'D'||}{||D'E'||} \cdot \frac{||E'F'||}{||F'G'||} \cdot \frac{||G'H'||}{||H'A'||} = \underline{\hspace{2cm}}.$$

The cross ratio and the circular product are both examples of a more general type of expression first described by Professor Howard Eves in his book *A Survey of Geometry* [19]. The definition below, which largely uses his original wording, does not explain the origin of the term "*h*-expression" (but we sometimes use the mnemonic "h" for "Howard").

Definition A ratio of a product of directed segments to another product of directed segments, where all the segments lie in one plane, is called an *h-expression* if it has the following properties:

1. If we replace $|AB|$ (or $\|AB\|$, if we use all positive values) by $A \times B$, and similarly treat all other segments appearing in the ratio, and then regard the resulting expression as an algebraic one in the letters A, B, etc., these letters can all be cancelled out.
2. If we replace $|AB|$ (or $\|AB\|$, if we use all positive values) by a letter, say ℓ, representing the line AB, and similarly treat all other segments appearing in the ratio, and then regard the resulting expression as an algebraic one in the letters ℓ, etc., these letters can all be cancelled out.

4. Which of the following expressions, according to Figure 11.13, is an *h*-expression:

$$\frac{|AB|}{|BC|} \frac{|CD|}{|DE|} \frac{|EF|}{|FG|} \frac{|GH|}{|HI|} \frac{|IJ|}{|JK|} \frac{|KL|}{|AL|} \quad \text{or} \quad \frac{|MN|}{|NO|} \frac{|OP|}{|PQ|} \frac{|QR|}{|RS|} \frac{|ST|}{|TM|} \quad ?$$

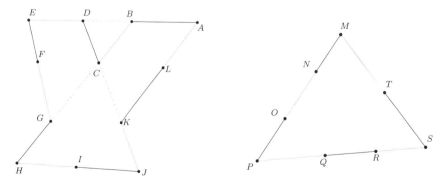

FIGURE 11.13: One of these figures gives an *h*-expression when we multiply the lengths of the dark line segments in the numerator and the lengths of the gray line segments in the denominator.

5. Is the h-expression from the previous problem positive or negative?

6. If we change the direction of one or more of the lines involved in an h-expression, how does the value of the h-expression change?

7. [T/F] The circular product from Figure 11.2.0 is an h-expression.

8. [T/F] The cross ratio is an h-expression.

The goal of questions 9–19 below is to prove the following:

Eves's Theorem: *The value of an h-expression is invariant under any projective transformation.*

9. Suppose O, P, and Q are three points in $\mathbb{E}^2$. Let $\angle POQ$ be the angle between 0 and π generated by the counterclockwise rotation of the line OP about O into the line OQ (therefore, $\angle POQ = 2\pi - \angle QOP$). Which of the following gives the area of the triangle $\triangle OPQ$?

(a) $\frac{1}{2}\|OP\| \cdot \|OQ\| \cdot |\cos(\angle POQ)|$ (c) $\frac{1}{2}\|OP\| \cdot \|OQ\|/|\cos(\angle POQ)|$

(b) $\frac{1}{2}\|OP\| \cdot \|OQ\| \cdot |\sin(\angle POQ)|$ (d) $\frac{1}{2}\|OP\| \cdot \|OQ\|/|\sin(\angle POQ)|$

10. Let $d_O(PQ)$ represent the distance from the point O to the line PQ (measured, usual, perpendicularly). Then one of the following statements is correct. Which one? And why?

(a) $d_O(PQ) \cdot \|PQ\| = \|OP\| \cdot \|OQ\| \cdot |\cos(\angle POQ)|$

(b) $d_O(PQ) \cdot \|PQ\| = \|OP\| \cdot \|OQ\| \cdot |\sin(\angle POQ)|$

(c) $d_O(PQ) \cdot \|PQ\| = \|OP\| \cdot \|OQ\|/|\cos(\angle POQ)|$

(d) $d_O(PQ) \cdot \|PQ\| = \|OP\| \cdot \|OQ\|/|\sin(\angle POQ)|$

When we prove a theorem as general as Eves's theorem, it helps to have a specific instance of the theorem in mind. So for problems 11–19 below, we will consider the h-expression

$$\mathcal{H} = \frac{\|AB\|}{\|BC\|} \cdot \frac{\|CD\|}{\|DE\|} \cdot \frac{\|EF\|}{\|FA\|},$$

and a perspective mapping $'$ from an ordinary center O as hinted at in Figure 11.14. Because every collineation is a composition of perspective collineations, proving Eves's theorem for perspective maps suffices. Of course, to fully prove the theorem, we would need to prove it for all h-expressions, not just for this one.

11. In Figure 11.14, we have labeled the points A and C. Add the labels B, D, E, F on the object triangle so that $\mathcal{H}$ (given above) forms an h-expression. Then label points A'–F' on the image triangle and O, the point of perspectivity.

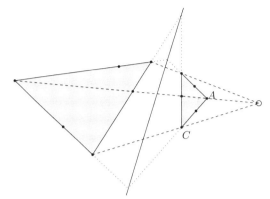

FIGURE 11.14: Triangles in perspective. (The shading is for artistic, not mathematical, purposes.)

12. Use your answer to question 10 above to replace $\|AB\|$ in the h-expression for $\mathcal{H}$ above. Do a similar replacement with all other lengths. Cancel distances where possible to simplify this version of the expression as much as possible.

$$\mathcal{H} = \frac{\|AB\|}{\|BC\|} \cdot \frac{\|CD\|}{\|DE\|} \cdot \frac{\|EF\|}{\|FA\|}$$

13. Would the cancellation and simplification you performed in the previous step work similarly for all other h-expressions? Why or why not?

14. Perform a similar replacement for the lengths $\|A'B'\|$, etc. in the expression for $\mathcal{H}'$. Again, cancel distances where possible to simplify this version of the expression.

$$\mathcal{H}' = \frac{\|A'B'\|}{\|B'C'\|} \cdot \frac{\|C'D'\|}{\|D'E'\|} \cdot \frac{\|E'F'\|}{\|F'A'\|}$$

15. If it is always true that $\angle OAB = \angle OA'B'$ (and similiarly for all other angles), then our the perspective invariance of $\mathcal{H}$ follows easily from comparing answers to questions 12 and 14. Are these angles pairwise equal in Figure 11.14?

16. Of course, Figure 11.14 shows an especially nice perspective map between two triangles, but we know that perspective maps can behave in nonintuitive ways instead. In Figure 11.15, we have labeled the points A, B, C, A', C', and E'.
    (a) Add the labels D, E, F on the object triangle so that $\mathcal{H}$ forms an h-expression.
    (b) Then locate O, the point of perspectivity.
    (c) Locate and label the remaining image points B', D', and F'.

17. If you got the previous answer correct, you will see that A and A' are on opposite sides of O, while B and B' are on the same side of O. Is it still true that $\angle OAB = \angle OA'B'$? (You might need to refer to how we measure angles, from the description in question 10.)

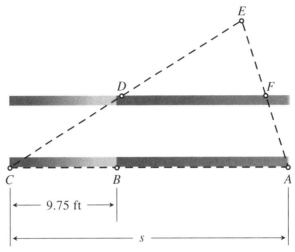

FIGURE 11.17: A view of the crash scene from above.

20. Explain why, in the plan view of Figure 11.17, the circular product depends only on the lengths of the wheelbase and skid marks, not on the specific location of the point E.

21. Use the perspective invariance of the circular product, together with measurements you make on the photograph of Figure 11.16, to determine the lengths of the skid marks.

22. Locate the vanishing point of the skid marks and road; call that V'. Use the cross ratio (AC, BV) to determine the length of the skid marks. How does your answer compare to the number you got in question 21?

THE CIRCULAR PRODUCT AND GEOMETRIC MEANS

Most people are familiar with the arithmetic mean: given a set of numbers $x_1, \ldots, x_n$, the arithmetic mean $(x_1 + x_2 + x_3 + \cdots + x_n)/n$ describes what you'd get, for example, if n people each put all their money into one pot, and then divided the contents of that pot equally among themselves. But what does the geometric mean $\sqrt[n]{x_1 \cdot x_2 \cdot x_3 \cdots x_n}$ mean? Here is one geometric interpretation.

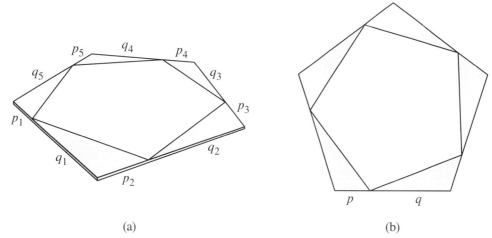

(a)   (b)

FIGURE 11.18: A rotated tile.

Suppose we have a regular n-gon inscribed in a larger regular n-gon, such as the decorative tile in Figure 11.18. Each vertex of the smaller n-gon divides an edge of the larger one into lengths p and q. Suppose we see the n-gon in perspective, as on the right side of Figure 11.18, with the lengths p, q transformed into lengths p_n, q_n.

23. Give an explanation of how Eves's theorem applied to Figure 11.18 gives us a visual interpretation of the geometric mean.

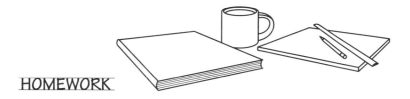

HOMEWORK

EXERCISES

(E) 11.2.1. Figure 11.19 shows a "photograph" (or some perspective image) of two squares; the small one is twisted so its four corners lie exactly on the four edges of the larger one. If the original large square is 1 m (100 cm) on each side, what is the length of the side of the smaller square?

(E) 11.2.2. Usually, mathematics does not let us prove a theorem by demonstrating one specific example for which the theorem holds. We may not, for example, prove that the sum of two odd integers is always an even integer merely by noting that $3 + 5 = 8$.

But examples are sometimes enough to prove theorems in the realm of projective geometry applied to perspective art. One of the truly lovely aspects of projective invariants is that they allow us to prove some theorems not in their full abstract generality, but rather in one specific and very easy-to-compute circumstance. Then we can wave our magic projective wand and say, "so this always works!"

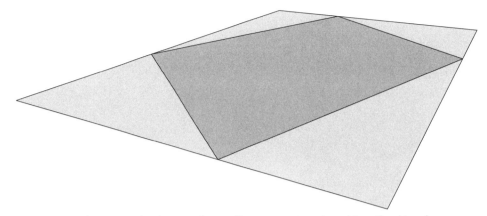

FIGURE 11.19: A perspective image of a small square, rotated, and inscribed in a larger square, for homework exercise Ⓔ 11.2.1.

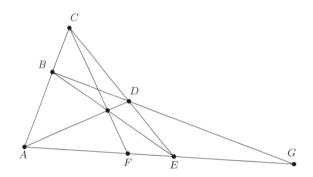

FIGURE 11.20: One possible diagram illustrating Ceva's and Menelaus's theorems.

(a) Prove Ceva's theorem by being clever (that is, by finding an instance where the theorem is obviously true) and using Eves's theorem to deduce therefore that the theorem is always true. Recall that Ceva's theorem says that, given a triangle $\triangle ACE$ with points B on AC, D on CE, and F on EA, with B, D, and F distinct from A, C, and E, the lines AD, CF, and EB are concurrent if and only if

$$\frac{|AB|}{|BC|} \cdot \frac{|CD|}{|DE|} \cdot \frac{|EF|}{|FA|} = 1.$$

(b) Do the same for Menelaus's theorem, which says that given a triangle $\triangle ACE$ with points B on AC, D on CE, and G on EA, with B, D, and G distinct from A, C, and E, the points B, D, and G are collinear if and only if

$$\frac{|AB|}{|BC|} \cdot \frac{|CD|}{|DE|} \cdot \frac{|EG|}{|GA|} = -1.$$

ART ASSIGNMENT

⚠ 11.2.1. Locate (and if possible, photograph) an interesting instance of an h-expression in the real world. Cross ratios are in many places; circular products are somewhat harder to find. Locating an instance of an h-expression that is neither of these requires real creativity!

PROOF/COUNTEREXAMPLE

P 11.2.1. Suppose we have a perspective mapping with center O, and that for some directed line ℓ we give ℓ' the induced direction from the perspective mapping: that is, if the ideal point of ℓ maps to a point outside of the segment $X'Y'$, then $|XY|$ has the same sign as $|X'Y'|$.

Suppose moreover that X and X' are on the same side of O (that is, O does not lie in the finite segment between X and X'. Prove that $\frac{|XY|}{|X'Y'|} > 0$ implies that Y and Y' also lie on the same side of O.

P 11.2.2. Adapt the arguments from this module to prove that an h-expression involving directed distances is invariant under a perspective collineation whose center is an ordinary point. (*Hint*: Start with a set of directions for the lines in the object, and use the induced directions for the lines in the image.)

FIGURE 11.3.0: Adding circles to a picture of a storefront. [For use with the AN ANGLE ON PERSPECTIVE: CASEY'S THEOREM module.]

11.3

An Angle on Perspective

Casey's Theorem

Overview When we draw a perspective picture or take a photograph, the angles in the picture don't generally match the angles of the objects from the real world. But we can nonetheless determine a real-world angle by drawing circles and lines in the picture plane and measuring a resulting angle, called the *Casey angle*.

---

1. Given four equally spaced, collinear points A, C, B, D (below), and overlapping circles with diameters AC and BD, what is the exact degree measure of the "Casey angle" $\angle BOC$? You should not need to measure; use geometry and logic.

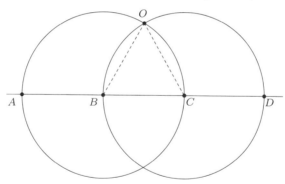

2. In the photograph in Figure 11.3.0, draw and measure the Casey angle: $\mathrm{m}(\angle B'PC') =$ _____.

   Note that although A', B', C', and D' are images of their respective points, the circles in Figure 11.3.0 are not *images of circles in the real world—they are just circles we drew on the photograph. Likewise, P is not the image of O.*

Definition Given four distinct collinear points, denoted in order by A, B, C, and D, let O be an intersection point of the circles with diameters AC and BD. Then the angle $\angle BOC$ is the *Casey angle* of those four points.

Note that, because there are two choices of O, there are two different possible Casey angles, differing only by sign. We will chose the angle $\theta \in (0, \frac{\pi}{2})$.

PROOF OF CASEY'S THEOREM

Casey's Theorem: *The Casey angle is a projective invariant.*

When we proved Eves's theorem, we converted an expression in lengths to an expression in angles. So it seems only fitting to prove Casey's theorem by describing an angle in terms of ratios of lengths.

Let A, B, C, D be distinct, collinear points appearing in that order, with Casey angle θ located at point O. Denote the angles by $\alpha = \angle AOB$, $\beta = \angle ABO$, $\gamma = \angle COD$, and $\delta = \angle CDO$, as in Figure 11.22.

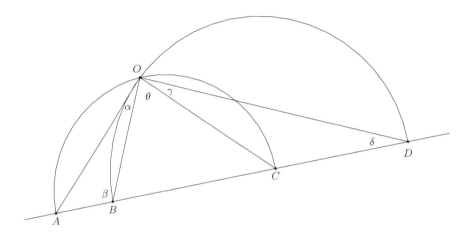

FIGURE 11.22: The Casey angle θ of the set of points A, B, C, D.

3. How are α and γ related to θ?
4. Which of the following quantites are equal to $\cos(\theta)$?

(a) $\sin(\alpha)$	(e) $\sin(\gamma)$	(i) $\sin(\alpha + \theta + \gamma)$
(b) $-\sin(\alpha)$	(f) $-\sin(\gamma)$	(j) $-\sin(\alpha + \theta + \gamma)$
(c) $\cos(\alpha)$	(g) $\cos(\gamma)$	(k) $\cos(\alpha + \theta + \gamma)$
(d) $-\cos(\alpha)$	(h) $-\cos(\gamma)$	(l) $-\cos(\alpha + \theta + \gamma)$

5. Use the law of sines and your answer to the previous question to determine a formula for the length $\|AB\|$ in terms of $\|OA\|$, β, and θ.
6. Determine a formula for $\|AB\|/\|BC\|$ in terms of $\|OA\|$, $\|OC\|$, and θ.
7. Determine a formula for the length $\|AD\|$ in terms of $\|OA\|$, δ, and θ.
8. Using the methods from the questions above, develop a formula for the cross ratio (AC, BD). Simplify as much as possible.
9. [T/F] The cross ratio is a projective invariant.
10. [T/F] The Casey angle is a projective invariant.

Application of Casey's theorem with vanishing points

Up to this point, we have assumed the four points A, B, C, D are ordinary points. But a very useful application of Casey's theorem follows when the four points are ideal. The connection between Casey angles and ideal points makes some intuitive sense: a Casey angle helps us to understand the angle between lines that connect to points; an ideal point help us understand the direction of a line.

But how do we make the connection between Casey angles and ideal points explicit? Figure 11.23 shows one possible attempt. Does this attempt work?

The figure shows an aerial photograph of (looking straight down at) a piece of land with two roads—Vine Street and Water Street—meeting at an angle of 125° (or 55°, depending on which angle we measure). In a happy coincidence, each street just happens to have a power line running exactly perpendicular to the road (the dotted lines).

One geometer theorizes that we can measure the angle between the roads by using the Casey angle in this manner: mark points A, B, C, D where the power lines and the road's dividing lines meet the northern edge of the map as shown in Figure 11.23. Draw the semicircles with diameter AC and BD, and let P be the intersection of these points.

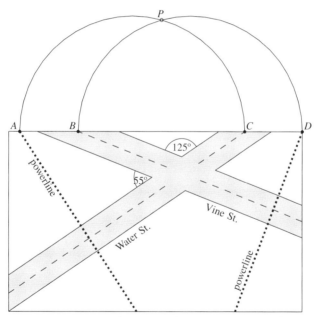

FIGURE 11.23: An aerial image with an attempt to apply Casey angles to one edge.

11. Vine Street and its power line meet at right angles. Is $\angle BPD = 90°$?
12. Water Street and its power line meet at right angles. Is $\angle APC = 90°$?
13. Vine Street and Water Street meet at 125° or 55°. Is $\angle BPC$ equal to either of these numbers? What about $\angle APD$?
14. Do A, B, C, D form a harmonic set? That is, do we have $\mathsf{H}(AC, BD)$?

Another geometer approaches the question differently, as in Figure 11.24. This geometer pretends that the photograph shows a plan view, with the top edge representing the picture plane. The geometer adds a viewer at a some point O, and then draws lines from

O parallel to the roads and power lines, labeling the intersection of these lines with the picture plane V, V', W, W' as in the figure.

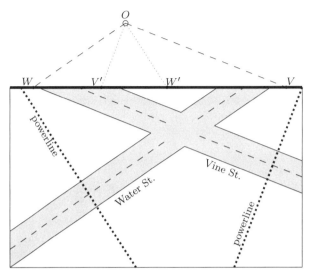

FIGURE 11.24: A top view interpretation of the aerial image.

15. Vine Street and its power line meet at right angles. Is $\angle V'OV = 90°$?

16. Water Street and its power line meet at right angles. Is $\angle WOW' = 90°$?

17. Do the four points form a harmonic set? That is, do we have $\mathsf{H}(WW', VV')$?

18. Vine Street and Water Street meet at 125° or 55°. Is $\angle VOW$ equal to either of these numbers? What about $\angle V'OW'$?

19. Given a picture plane ω, and given a figure containing at least two lines f and g in a plane α perpendicular to ω. Denote by ℓ_α the vanishing line of α in ω.
    (a) What can we say about the relationship between the viewing target T and the line ℓ_α?

    (b) Is the angle between f' and g' the same as the angle between f and g?

    (c) The plane ω contains a point O with $\|OT\|$ being the viewing distance and OT perpendicular to ℓ_α. Explain how we can use O, f', and g' to determine the angle between f and g.

20. Figure 11.25 shows part of what an observer would see, according to the plan view in Figure 11.24. We have drawn the image of Vine Street, but not of Water Street or of either of the power lines. We have, however, marked the viewing target T, the viewing distance $||OT||$, and the intersection of Water Street and the power lines with the picture plane. (Points A and D match the points in Figure 11.23.) Use Figure 11.24, a protractor, and your answer to question 19 to finish the picture.

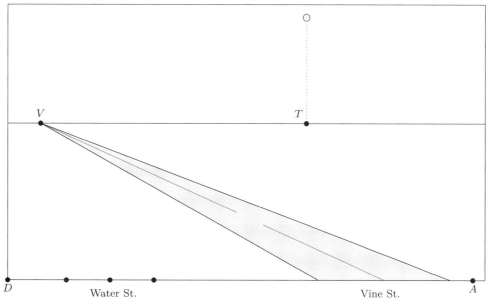

FIGURE 11.25: Draw correct perspective images of Water Street and the power lines.

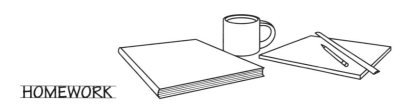

HOMEWORK

EXERCISES

ⓔ 11.3.1. What is the angle formed in the space between the book and the notepad whose images appear in the homework icon above?

ⓔ 11.3.2. Assume the two squares whose images appear in Figure 11.26 lie in a plane that is perpendicular to the picture plane; the line in the figure is the horizon line. What is the angle between the squares? Use your answer to determine the ratio between the sizes of the two squares.

ⓔ 11.3.3. Draw the perspective image of an equilateral triangle that lies perpendicular to the picture plane. Assume the viewing target is in the center of the paper and the viewing distance is 4 inches.

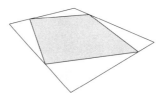

FIGURE 11.26: A perspective image of a small square, rotated, and inscribed in a larger square, for homework exercise Ⓔ 11.3.2.

ART ASSIGNMENT

⚠ 11.3.1. Draw a floor tiled by hexagonal tiles, with none of the lines in the tiles parallel to the picture plane, and none of the lines in the tiles vanishing at the viewing target. Draw at least four tiles across and three tiles deep. Indicate the viewing target and viewing distance on your drawing.

If this hexagonal grid were on the floor of a normal room, the room would be rectangular. Draw the image of a large rectangle over your grid; some hexagons and parts of others might stick out of the edges of the rectangle, but do not make the rectangle so large that you leave "holes" in your floor.

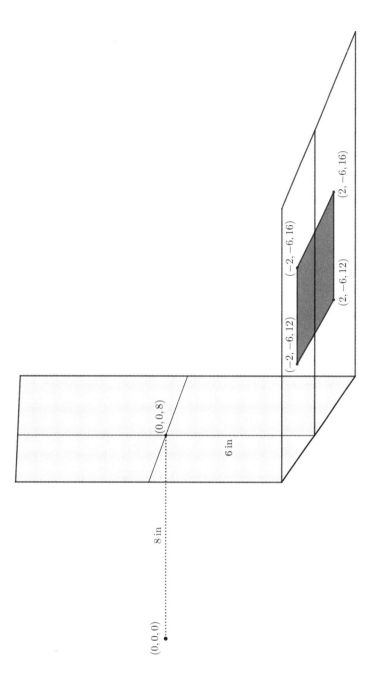

FIGURE 12.1.0: A picture of coordinates. [For use with the EUCLIDEAN GEOMETRY ENHANCED WITH ALGEBRA module.]

12

Coordinate Geometry

12.1 Euclidean Geometry Enhanced with Algebra

Overview In this module, we connect the making of perspective pictures with a review of three-dimensional Cartesian coordinate space. In other words, we do perspective "by the numbers."

In Figure 12.1, we draw the point $P = (3, 4, 6)$ in three-dimensional space. To indicate the location of P, we go 3 units to the right in the x direction, 4 units up in the y direction, and 6 units forward in the z direction. This way of describing space is called a *Cartesian coordinate system*. (You may be used to a different orientation from multivariable calculus; this "left-handed" version of a coordinate system will allow us to label a picture plane using familiar xy-coordinates.)

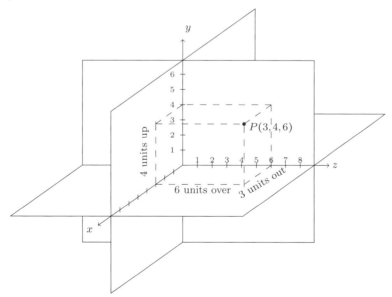

FIGURE 12.1: A point floating in space.

1. The side and top views in Figure 12.6 show the line of sight between the eye of the viewer and a point on the object, with the picture plane a distance d from the viewer. Additionally, we have superimposed a Cartesian coordinate system with the eye at $(0, 0, 0)$. Fill in the blanks below with the words "in front of," "behind," "above," "below," "to the left of," and "to the right of."

(a) Positive x-values represent the portion of space that is _____ the viewer's eye.

(b) Negative x-values represent the portion of space that is _____ the viewer's eye.

(c) Positive y-values represent the portion of space that is _____ the viewer's eye.

(d) Negative y-values represent the portion of space that is _____ the viewer's eye.

(e) Positive z-values represent the portion of space that is _____ the viewer's eye.

(f) Negative z-values represent the portion of space that is _____ the viewer's eye.

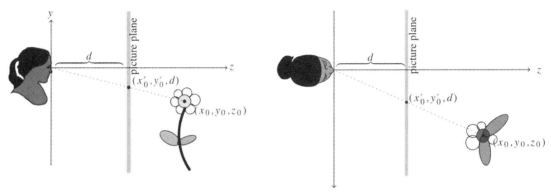

FIGURE 12.2: A side view and top view with superimposed axes. In the side view, the x-axis is coming directly toward us; in the top view, the y-axis is pointing toward us.

2. If the picture plane is at $z = d$, then a point $P = (x_0, y_0, z_0)$ has an image at a point $P' = (x'_0, y'_0, d)$.

(a) Draw and label the lengths x_0, x'_0, and z_0 in the top view of Figure 12.6.

(b) Draw and label the lengths y_0, y'_0, and z_0 in the side view of Figure 12.6.

(c) Use similar triangles to determine formulas for x'_0 and y'_0 in terms of x_0, y_0, z_0, and d.

• $x'_0 =$ _____ $y'_0 =$ _____

In the exercise and questions that follow, we will do a physical demonstration of your algebraic formula as in Figure 12.1.0. You will need

- a "window": either a small sheet of Plexiglass (roughly the same size as a piece of paper) or a sheet of transparency paper,
- an erasable marker that allows you to draw on your window,
- a "floor": a blank sheet of paper,
- a pencil that allows you to draw on your floor, and
- a ruler.

Eventually, you will have you place your eye at $(0, 0, 0)$, the window will be the plane $z = 8$, and the coordinates of a square on the ground will be $A = (-2, -6, 12)$, $B = (2, -6, 12)$, $C = (-2, -6, 16)$, and $D = (-2, -6, 16)$.

3. Set up the window and draw the x- and y-axes on the $z = 8$ plane (the Plexiglass or transparency) so that the viewing target $(0, 0, 8)$ is 6 units from the bottom.
4. Place a piece of paper along the bottom edge of the window and draw the z-axis on the $y = -6$ plane (the ground).
5. On the ground, draw the points A, B, C, and D of the square and connect them to draw the edges. (*Note*: Since the window is at $z = 8$, the point $A = (-2, -6, 12)$ has z-coordinate only 4 units from the window.)
6. Use your answer to question 2 to determine the coordinates of these image points:
   (a) $A' = $ _____
   (b) $B' = $ _____
   (c) $C' = $ _____
   (d) $D' = $ _____
7. Draw the image points on the Plexiglass and connect them to make the edges of a quadrangle.

8. Place your eye at the appropriate location, and verify that the image points do indeed line up with the actual points in the real world!
9. Extend the edges of the image. Where is/are the vanishing points (if any)?

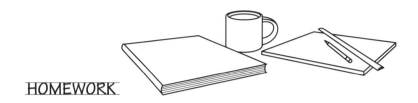

HOMEWORK

EXERCISES

ⓔ 12.1.1. On the axes below, draw the following points. You do not need to draw the planes, but draw some of the dashed lines that indicate distance from axes, as in Figure 12.1.

(a) $(2, 3, 4)$
(b) $(2, 3, -4)$
(c) $(2, -3, 4)$
(d) $(-2, 3, 4)$

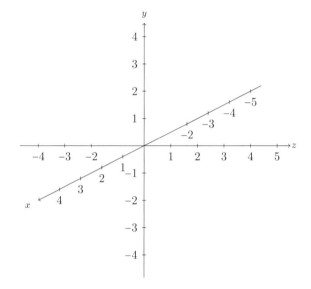

ⓔ 12.1.2. The axes below go from -4 to 4 in each direction; in other words, we have drawn an $8 \times 8 \times 8$ cube. Draw the following planes as they intersect this cube.

(a) $x = 2$
(b) $y = -3$
(c) $z = -1$

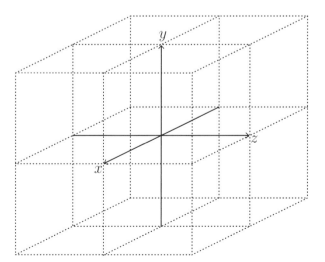

(E) 12.1.3. A *parallelepiped* is a three-dimensional polyhedron whose eight faces are all parallelograms. That is, each edge of a parallelepiped is parallel to exactly three other edges.

(a) Open up a computer spreadsheet, such as Excel (see Figure 12.3), Google Docs (Figure 12.4), or GEOGEBRA. In this spreadsheet, input the points of the parallelepiped that has the eight vertices

$$A = (1, -1, 2), \qquad B = (1, -1, 5), \qquad C = (3, -1, 7), \qquad D = (3, -1, 4)$$

$$E = (1, -2, 2), \qquad F = (1, -2, 5), \qquad G = (3, -2, 7), \qquad H = (3, -2, 4).$$

Input these vertices in three columns, one for each of the x, y, and z coordinates.

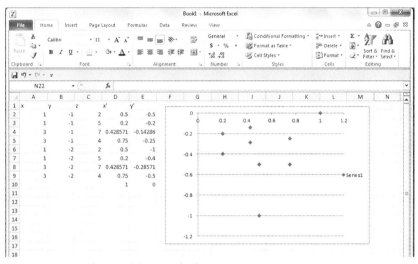

FIGURE 12.3: Excel spreadsheet with chart.

FIGURE 12.4: Google Docs spreadsheet with chart.

(b) Using your formula for x' and y' with $d = 1$, input x' and y' in two other columns, respectively.

(c) Add one more point under the x' and y' columns: $x' = 1$ and $y' = 0$.

(d) Highlight the columns for x' and y' and make a scatterplot. Make sure the units for your axes are roughly the same, otherwise your picture may look distorted (see Figure 12.3).

(e) Print the image and connect the pairs of dots that ought to be connected (that is, the images of edges but not the images of diagonals).

(f) In a parallelepiped, there are three sets of parallel lines. Highlight the images of these sets in three different colors.

(g) Which set of lines was parallel to the picture plane? Explain why the image of the set remained parallel.

(h) In Figure 12.5, we have started drawing the top view of the parallelepiped. Label the axes, add the viewer, and add the picture plane. Then use this top view to explain why, in your spreadsheet image, one set of edges of the parallelepiped has images that vanish at $(0, 0, 1)$ and another set vanishes at $(1, 0, 1)$.

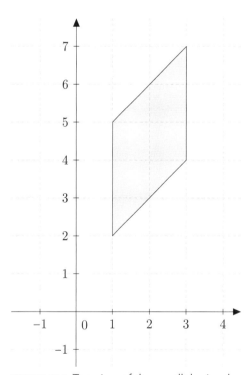

FIGURE 12.5: Top view of the parallelepiped.

12.2 Introduction to Homogeneous Coordinates

Overview Cartesian coordinates allow us to use algebra to describe geometric ideas in Euclidean space, but perspective pictures don't always behave like their real-world counterparts (for example, the images of parallel railroad tracks usually meet at a point). This module introduces a useful kind of coordinate system for perspective and projective images, called *homogeneous coordinates*.

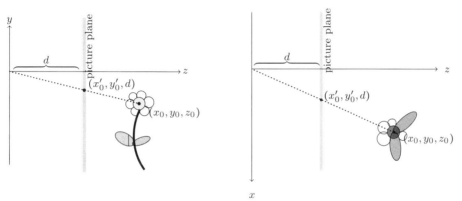

FIGURE 12.6: A side view and top view with superimposed axes. In the side view, the x-axis is coming directly toward us; in the top view, the y-axis is pointing toward us.

1. As a quick review of Cartesian coordinates, use similar triangles in Figure 12.6 to determine formulas for x_0' and y_0' in terms of x_0, y_0, z_0, and d.
   - $x_0' = $ _____ $y_0' = $ _____

2. Let the point on the flower in Figure 12.6 be $F = (2, -4, 4)$. If the viewing distance is $d = 1$, what are the coordinates of the image F'?

3. Give the coordinates of three other points that have the same image F', that is, three other points that are on the same line of sight between the eye at $O = (0, 0, 0)$ and $F = (2, -4, 4)$.

4. What is the relationship between the points you found? What is the general pattern of *all* points that project to F'?

We call the point $O = (0, 0, 0)$ the *center* of the perspective mapping. In the IMAGE OF A LINE module, we describe three physical models for making perspective images:

- an artist looking through a window (in which the image plane lies between the object and the center of the perspective mapping),
- a light source casting a shadow (in which the object lies between the image and the center of the perspective mapping), and
- a pinhole camera (in which the center of the perspective mapping lies between the object and the image).

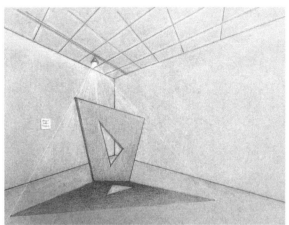

FIGURE 12.7: Shadows and pinhole cameras.
Study for the cover of our book by Fumiko Futamura; (Camera Obscura): Wellcome Collection

5. The points $C = (-2, 4, -4)$ and $S = (\frac{1}{4}, -\frac{1}{2}, \frac{1}{2})$ lie on the line OF, that is, both points could be considered to have F' as an image. Which of these points acts like an object that casts a shadow with O as the light? Which of these points acts like an object that is viewed by a pin-hole camera with the hole at O?

What we have said so far is that, from the point O, almost every point on the line OF—the sole exception being O itself—maps to the same point F'. Mathematicians say that this gives us an *equivalence*, which we denote by the symbol "$\sim$". For example, we write

$$(2, -4, 4) \sim \left(\frac{1}{4}, -\frac{1}{2}, \frac{1}{2} \right).$$

Moreover, we can think of the set of all points equivalent to F as an *equivalence class* of points $\langle F \rangle$. That is, $\langle F \rangle$ is the set of all points on the line OF, with the exception of the point O.

6. Give a reason from perspective viewing why we would want to exclude the point $(0, 0, 0)$ from the equivalence class $\langle F \rangle$.
7. Is it true or false that $\langle F \rangle = \langle F' \rangle$?

Definition We will define a *homogeneous point* $(x : y : z)$, or equivalently a *point in the projective plane*, to be the equivalence class of points in $\mathbb{R}^3$ minus the origin under the equivalence $\sim$, so that

$$(cx : cy : cz) = (x : y : z)$$

for $c \neq 0$.

Note that we use slightly different notation for two different kinds of objects. For the real point $(x, y, z) \in \mathbb{R}^3$, we write $(cx, cy, cz) \sim (x, y, z)$; for the homogeneous point $(x : y : z)$ in the projective plane, we write $(cx : cy : cz) = (x : y : z)$.

There is a close connection between the projective plane and the extended Euclidean plane $\mathbb{E}^2$. Where possible, we will use the fact that if $z \neq 0$, then we consider $(zx : zy : z)$ and $(x : y : 1)$ to be the same homogeneous point. That is, we will project real points to the picture plane $z = 1$; so we could think of these as corresponding to ordinary points in $\mathbb{E}^2$. As the examples in questions 10 and 11 below show, for this projection, homogeneous points with $z = 0$ correspond to ideal points in $\mathbb{E}^2$.

8. Consider the following set of points in real space:

$$\{(0, -1, 1), (0, -1, 2), (0, -1, 3), \ldots\} = \{(0, -1, n)\}_{n \in \mathbb{N}}.$$

We could think of these as points straight ahead of ($x = 0$) but below ($y = -1$) the viewer, equally spaced along a line extending into the distance. For example, possibly these are pebbles on a sidewalk heading away from us, as in Figure 12.8.

(a) If we think of these instead as homogeneous coordinates, rewrite these points where possible with $z = 1$.

(b) Considered as homogeneous points, these points converge to which homogeneous point as $n \to \infty$? (In other words, which line in $\mathbb{R}^3$ describes the correct equivalence class?)

(c) How does this answer match our previous perspective notions?

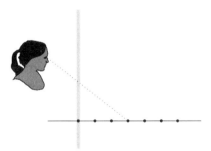

FIGURE 12.8: Side view showing a viewer facing a pebble-strewn road.

9. Consider the following set of points in real space:

$$\{(1, -1, 1), (1, -1, 2), (1, -1, 3), \ldots\} = \{(1, -1, n)\}_{n \in \mathbb{N}}.$$

We could think of these as points slightly to our right ($x = 1$) and below ($y = -1$) the viewer, equally spaced along a line extending into the distance. For example, possibly these are the railroad ties on the second track heading away from us.

(a) Rewrite these points as homogeneous points with $z = 1$.

(b) Considered as homogeneous points, these points converge to which homogeneous point as $n \to \infty$? (In other words, which line in $\mathbb{R}^3$ describes the correct equivalence class?)

(c) How does this answer match our previous perspective notions?

10. Consider the following set of points in homogeneous coordinates:

$$\{(1:-1:1), (2:-1:1), (3:-1:1), \dots\} = \{(n:-1:1)\}_{n \in \mathbb{N}}.$$

(a) Give a physical interpretation for what these points might represent.

(b) These points converge to which homogeneous point as $n \to \infty$? (In other words, which line in $\mathbb{R}^3$ describes the correct equivalence class?)

11. Consider another set of points in homogeneous coordinates:

$$\{(1:2:1), (2:2:1), (3:2:1), \dots\} = \{(n:2:1)\}_{n \in \mathbb{N}}.$$

(a) Give a physical interpretation for what these points might represent.

(b) These points converge to which homogeneous point as $n \to \infty$? (In other words, which line in $\mathbb{R}^3$ describes the correct equivalence class?)

12. *Circle the correct phrase(s):* A point in [real space / the projective plane] represents a line through the origin in [real space / the projective plane] .

13. *By analogy with question 12, fill in the blanks*: A line in the projective plane represents a _____ through the origin in _____ .

Let us build on our answer to question 13, and consider how we can use planes through the origin to give us projective lines. A plane $\alpha \subset \mathbb{R}^3$ has equation $ax + by + cz = d$.

14. If the plane passes through the origin, what (if anything) can we say about the constants a, b, c, and d?

15. Suppose moreover that the two points $(0, 2, 1) \in \mathbb{R}^3$ and $(1, 2, 1) \in \mathbb{R}^3$ lie on α. (Since three points in $\mathbb{R}^3$ determine a plane, these two points together with $(0, 0, 0)$ completely determine α. What (if anything) can we say about the constants a, b, c, and d?

16. [T/F] Any point $(x, 2, 1)$ with $x \in \mathbb{R}$ lies on α.

Definition We will define a *homogeneous line* $[a:b:c]$, equivalently a *line in the projective plane*, to be the equivalence class of lines that lie in a real plane with equation $ax + by + cz = 0$. It follows that $[a:b:c] = [wa:wb:wc]$ for any nonzero $w \in \mathbb{R}$.

17. Write the homogeneous line coming from plane $y - 2z = 0$ in the form $[a:b:c]$.

18. Verify that the homogeneous point at infinity that you got in question 11(b) lies on this homogeneous line.

19. Determine the homogeneous line $[a:b:c]$ that contains the points

$$\{(1:-1:1), (2:-1:1), (3:-1:1), \dots\}$$

described in 10.

20. Given a homogeneous point $(x:y:z)$ and a homogeneous line $[a:b:c]$, how can we determine if the point is incident with the line?

21. Given two distinct homogeneous points $(x:y:z)$ and $(x':y':z')$, how do we create the homogeneous line determined by these points?

22. Given two distinct homogeneous lines $[a:b:c]$ and $[a':b':c']$, how do we create the homogeneous point determined by these lines?

Supplementary material: Linear algebra

This section is for students who have taken Linear Algebra or Multivariable Calculus. If we understand the language of vectors, we can use ideas like *dot product, cross product, linear independence* and *orthogonality* to determine intersections and incidence. Recall the following facts regarding vectors in $\mathbb{R}^3$:

- A plane through the origin is determined by a vector orthogonal to the plane; that is, if a nonzero vector (a, b, c) is orthogonal to a plane through the origin, then the equation of the plane is $ax + by + cz = 0$.
- Two vectors are orthogonal if and only if their dot product equals zero.
- The cross product of two linearly independent vectors gives a nonzero vector orthogonal to both.
- Three vectors are linearly independent if and only if the determinant of the matrix with these vectors as columns is nonzero.

Answer the following questions about homogeneous points and lines.

23. Given a point $(x:y:z)$ and a line $[a:b:c]$, how can we use the language of vectors to determine if the point is incident with the line?

24. Given two distinct points $(x : y : z)$ and $(x' : y' : z')$, how do we use the language of vectors to create the line determined by these points?

25. Given two distinct lines $[a : b : c]$ and $[a' : b' : c']$, how do we use the language of vectors to create the point determined by these lines?

26. Given three distinct points, how can we test to determine whether they are collinear?

27. Dually, given three distinct lines, how can we test to determine if they are concurrent?

FIGURE 13.0: A template for Möbius shorts. [For use with the THE SHAPE OF EXTENDED SPACE module.]

13

The Shape of Extended Space

A twisted view of perspective

Overview We have asked many questions along the line of, "What does such-and-such an object look like when we project it onto $\mathbb{E}^2$?" But we have not yet looked at the question of "What does $\mathbb{E}^2$ itself look like?" In this module, we work toward getting an overview of the shape (in the topological sense) of the extended Euclidean plane.

We will begin by exploring "surfaces with boundary." A *surface* is a shape that is made up of two dimensions everywhere, with the possible exception of the boundary (that is, the edge). Although the surface itself is two-dimensional, it might be a subset of a higher-dimensional space: for example, the surface of the earth is two-dimensional, but it exists in a three-dimensional space.

1. Without actually making these objects yourself (yet), imagine you have three $2'' \times 11''$ strips of paper labeled as in Figure 13.1, and that, for each strip of paper, you join together its two dotted edges so that the A's meet and the B's meet. If you join the ends properly, the black shapes will join to form triangles, not parallelograms.
   Match the resulting shapes with the images in Figure 13.2.

In terms of the aspects of geometry that require measuring distances, the long, skinny object is different from the other two objects. But we are going to care about the *topology* of the surfaces, that is, properties of shape that do not depend on distance.

2. Looking at these shapes topologically, two of the surfaces are *cylinders* and the third is not; the third is a shape called a *Möbius band*. Which of the surfaces in Figures 13.1–13.2 is the Möbius band?

Make your own Möbius band. Since we will be drawing on and cutting the band, be sure to tape the band together securely; see Figure 13.3.

3. If you think of the Möbius band as a road, you could draw the dividing line down the middle of this road. In Figure 13.1, the dashed arrows are a start to drawing this line. Continue this line on your own Möbius band until you get back to the where you started drawing. What is unusual about drawing this line?

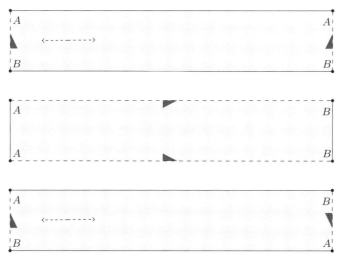

FIGURE 13.1: Three strips; join the dotted edges to form another shape.

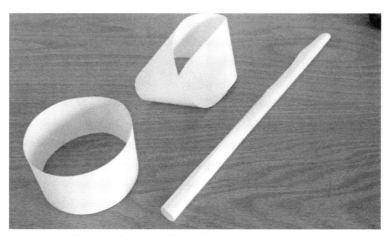

FIGURE 13.2: Three surfaces with boundary; match these with the strips in Figure 13.1.

FIGURE 13.3: Good taping technique (on the left) joins the entire edges; poor taping technique (on the right) allows corners to flap loosely.

4. What do you think will happen if you cut the Möbius band along that line? Try it! (Don't cut in from the edge; just cut along that line.)

5. What happens if you repeat this process: that is, you draw a line down the middle of the halved Möbius band, and cut along that?

6. Start over with a new Möbius band, properly taped. What happens if you cut the Möbius band into thirds?

OPTIONAL ACTIVITY_____

The Möbius band is a familiar object to most mathematicians. Somewhat less commonly known, but also incredibly fun to play with, are Möbius shorts, first described by Gourmalin [9, p. 47] and introduced to U.S. mathematicians by Ralph Boas [7]. A two-strip template appears in Figure 13.0; a completed version appears in Figure 13.4.

FIGURE 13.4: Möbius shorts, assembled.

7. Make your own set of Möbius shorts. Be sure to tape the three different edge-joins securely; you might want to have tape on both sides.

8. Confirm that this surface-with-boundary is one sided, like the Möbius band.

9. What happens if you cut the Möbius shorts "in half"? To ensure you only cut in the middle of the strips and not through the edges, you might first want to draw lines that begin and end at the T-shaped arrows.

10. Extra-optional: Once you've cut the shorts into their new shape, can you put them back together?

What does a Möbius band have to do with $\mathbb{E}^2$? Because in topology we often disregard distances, imagine that we can "shrink" $\mathbb{E}^2$ to a square as in Figure 13.5. The line ℓ has an ideal point $P_{[\![\ell]\!]}$, which appears in Figure 13.5 at two edges of the square; we could say the same for the line k. But notice that if we join the two images of these points together, we get a Möbius band lying in $\mathbb{E}^2$!

SURFACES WITHOUT BOUNDARY

The top row of Figure 13.6 shows how to construct several surfaces by joining the edges of squares. In surfaces 1–3, we join the opposite edges so that the colored darts form triangles; in surface 4, we gather all edge points into just one point. We'll assume these squares

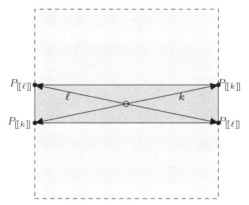

FIGURE 13.5: A sketch of how we can see a Möbius band in $\mathbb{E}^2$.

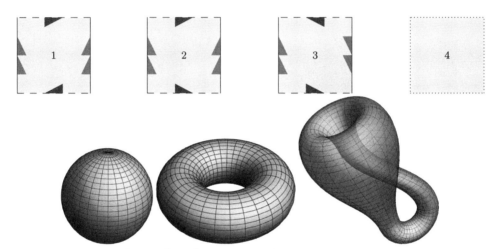

FIGURE 13.6: Four surfaces for questions 11–15.

are not made of paper, but rather some very stretchable substance (because in topology, distance is not as important as other properties of shape).

11. Which of the four surfaces in the top row forms a sphere?
12. Which of the four surfaces in the top row forms a torus (second shape in the bottom row)?
13. Which of the four surfaces in the top row forms a Klein bottle (third shape in the bottom row)?
14. Which of the four surfaces in the top row forms $\mathbb{E}^2$ (not pictured in the bottom row)?
15. Which of these four surfaces are non-orientable? (That is, which, like the Möbius band and Möbius shorts, allow us to draw on "both sides" of the object without lifting our pencil?)

Note that if we draw the Möbius band on a two-dimensional piece of paper, the image of the boundary always has a "self-intersection," even though the actual boundary in three-dimensional space does not. In the same way, with some of the surfaces in Figure 13.6, if we *embed* them in three-dimensional space, the embeddings will have self-intersections,

FIGURE 13.7: Any drawing of the Möbius band in the plane will have a self-intersection where the images of the edges cross. But a Möbius band sitting in space need not have self-intersections.

even though the surfaces themselves do not. (If we could embed them in four-dimensional space, we could avoid the self-intersections.)

ANOTHER VIEW: THE HALF SPHERE MODEL

Here is another way we can see $\mathbb{E}^2$. In creating the Möbius strip, we identified points on one pair of opposite edges; in question 14, we described $\mathbb{E}^2$ by identifying points on both pairs of opposite edges. For each of these constructions, we chose pairs of points on the *boundary* that we could join with one another to become one point.

Now let's do the same thing to a sphere S^2, though in this example *every* point will have a matching point. By identifying antipodal (that is, opposite) points, we get another topological model for the projective plane. To understand why and how this model describes $\mathbb{E}^2$, consider Figure 13.8. In this figure, we show that each point on the upper hemisphere is identified with its antipodal twin on the lower hemisphere (hence the points on the sphere are labeled, for example, A and A). We can map each of these (pairs of) points by a line through the center of a sphere to a point (say, A') on a horizontal plane tangent to the sphere. We can think of this plane as our usual version of $\mathbb{E}^2$ (in ways we explore in the

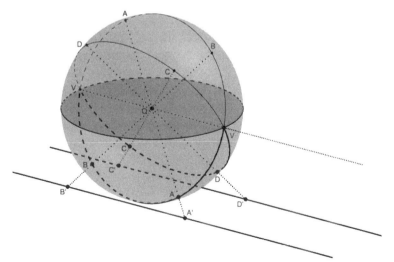

FIGURE 13.8: We identify antipodal points to get a topological model of $\mathbb{E}^2$.

questions below), and so the sphere with antipodal points identified is a topological way of seeing all the points of $\mathbb{E}^2$ in a compact space.

16. If the tangent plane is extended to $\mathbb{E}^2$, what part of the sphere maps to the ideal line?

17. What kind of subset of the sphere maps to an ordinary line in $\mathbb{E}^2$?

18. In Figure 13.8, we show a pair of parallel lines in the tangent plane. Describe the pencil of "lines" on the sphere that map to the full pencil of lines parallel to these lines.

SO WHAT DOES $\mathbb{E}^2$ ACTUALLY LOOK LIKE?

Getting an understanding of the shape of $\mathbb{E}^2$ is a difficult task, not because we don't know the shape but because the pictures of $\mathbb{E}^2$ are so twisted (literally) that following them doesn't help us "see" the space well. For example, Figure 13.9 shows three different well-known embeddings of $\mathbb{E}^2$ in $\mathbb{R}^3$.

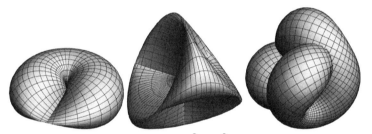

FIGURE 13.9: Three embeddings of $\mathbb{E}^2$ in $\mathbb{R}^3$: a *cross-cap, Roman's surface*, and *Boy's surface*.

To try to explore this unusual space, we could simplify the question by cutting some pieces out of $\mathbb{E}^2$ and studying only the part that remains. In Figure 13.10 we cut off the corners and retain the cross in the middle.

19. We could think of this Möbius cross as $\mathbb{E}^2$ with some holes cut out. How many holes does Figure 13.10 show in $\mathbb{E}^2$?
20. Create the Möbius cross. You can either cut this from a single sheet of paper or make two Möbius bands and tape them together at right angles.
21. What happens when you cut this Möbius cross in half?
22. Compare your answer to the previous question with those of other people in the class. Do you all get the same answer?
23. What shape do we get (topologically speaking) if we cut just one hole in $\mathbb{E}^2$?

FIGURE 13.10: A Möbius cross within $\mathbb{E}^2$.

Want to learn more about this subject?

August Ferdinand Möbius (1790–1868) worked as an astronomer, but is remembered mostly for his contributions to theoretical mathematics. Although a contemporary mathematician, Johann Benedict Listing (1808–1882), published the construction of the Möbius band four years before Möbius himself, it was Möbius who developed the working definition of *non-orientability* still in use today [20].

It is no coincidence that the projective plane happens to contain a subset that bears the name of Möbius. In the area of projective geometry, he made several significant contributions, including the notion of duality between points and planes in three-dimensional projective space; Möbius transformations of the complex plane; and, even more significantly for perspective art, the development of homogoneous coordinates to describe general projective spaces. Jeremy Gray writes in the *Princeton Companion to Mathematics*, "the simplicity and generality of Möbius's methods were important in establishing projective geometry as a rigorous mainstream subject" [26].

For more information on the subject of topology, there are many good introductory textbooks (for example, [5] and [49]). For more on the mathematics of surfaces, see, for example, *A topological Picturebook* [21] or *Surfaces* [28]. And for a lively, often humorous look at applications of Möbius bands, including the source for the two homework exercises below, see [43].

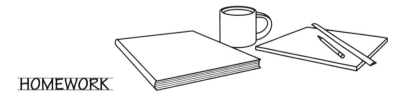

HOMEWORK

EXERCISES

Ⓔ 13.1. The four color theorem famously asserts that every map of the plane can be colored with four or fewer colors, in such a way that no two regions sharing the same color also share an edge. But four colors do not suffice for the Projective Plane $\mathbb{E}^2$. Create a map on $\mathbb{E}^2$ that requires six different colors (that is, any coloring of your map with five colors will necessarily have a pair of adjacent regions of the same color).

(E) 13.2. It is possible to connect four points in the plane with segments or curves so that every point is connected to every other point, but no two segements or curves intersect. If we have five points in the plane, though, we cannot connect all points pairwise without having the connecting segments cross each other.

But the Möbius band gives us more "room." Connect the six points in the Möbius band below in such a way that every pair of points is connected and such that curves have no crossings.

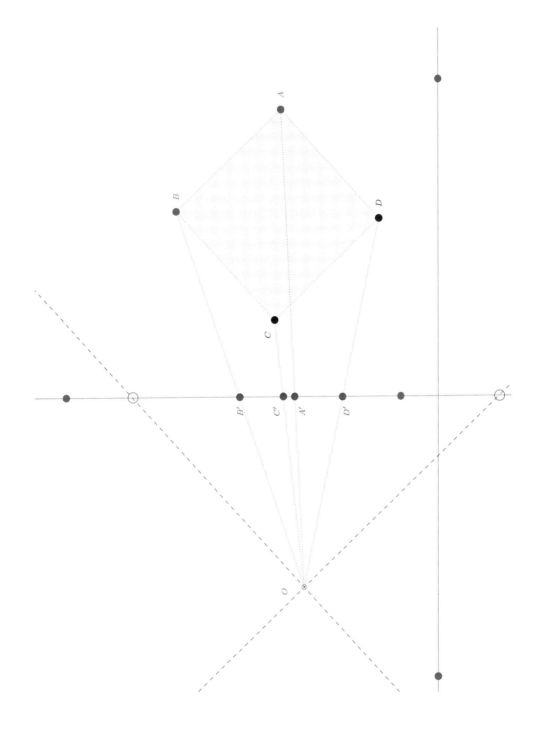

FIGURE G.0: A GEOGEBRA top view or side view for constructing the image of a square. [For use with the INTRODUCTION TO GEOGEBRA module.]

Appendix G

Introduction to GeoGebra

Overview In this module, we will first get an introduction to using GeoGebra; then we will use our knowledge to review top views and side views. Both GeoGebra and the top and side views will help us in the DYNAMIC CUBES AND VIEWING DISTANCE module.

Introduction to GeoGebra

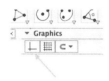

Open GeoGebra. You will want to turn off the axes. You can do this under View or by clicking the axis icon under Graphics. You might at times want your drawings uncluttered by labels; you can set this preference under Options ... Labeling, selecting No New Objects. Alternately, you can give an object a label of your choosing if you right-click on the object ("control" click, on a Mac), and choose Rename.

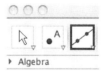

Select the Line tool and use this to make two different lines. Then choose the Point tool to make a point that is not on either of those lines. Click on the Move tool (the button with the Move), and note how you can move each object around independently of the others.

> The points are examples of what Venema [52] calls *free objects*; the lines are *dependent objects*. To paraphrase his description, the points are "free" in the sense that there is no restriction on where we can move them with the Move tool; the lines are "dependent" because we defined them through the placement of the points. If you delete one of the points, the line disappears; if you delete the line, the points remain.

Now we're going to change the color of the lines. While the Move tool is selected, click on one line and go to the Color icon under Graphics. Color one line red and the other line blue. You can change the color of the points on the line so they match, if you like. You may also change the color, shape, and size of the free-floating point, by clicking on the point and using the icons under Graphics.

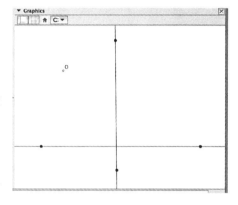

When you're at this stage, your diagram might look something like the one to the right. We'll call the free-floating point O, the roughly horizontal line r, and the roughly vertical line b. Note that this figure looks a bit like a top view or side view. For example, we could consider b to be like a picture plane, r, not a to be a road in the real world, and O to be the eye looking through the picture plane.

Where is the vanishing point of the r road line in the b picture plane? To determine the location, we'll need to select the Parallel Line tool. (You'll probably have to click the bottom corner of the Perpendicular Line tool, get a drop-down menu, and change to the Parallel Line tool).

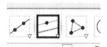

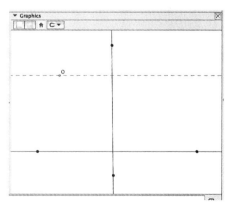

When you've selected the Parallel Line tool, click first on the point O and then on r. This diagram changed the parallel line to a dotted line, but you don't have to.

Then select the Intersection tool in the drop-down menu under the Point tool. Click on both the *b* picture plane line and the parallel line to get your point of intersection. We name this point 'V' in the subsequent diagrams. (*Question: Why is this an appropriate name?*)

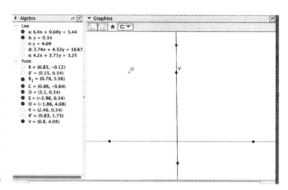

You don't need to see the parallel line anymore. If you *delete* that line, you'll lose the point *V* as well, but you can *hide* the line by clicking on the radio button (Line c in the version to the right) under the algebra list. You can also select the line and press Ctrl+G to hide the line. When you're done, you should have something like the figure to the right. Note that if you go back to the Move key, you can't move *V*, but you can move the original objects … and when you do move them, *V* moves to the correct place!

Here are some questions to play with.

1. What happens to *V* when you move *O*? How does your answer relate to the window taping exercise?

2. If you turn *r* in a circle around a fixed point (for example, if *r* were the spoke of a Ferris wheel), what does *V* do? How would we interpret this situation in the example of the eye, picture plane, and road?

3. What happens to V r red and b are parallel? Again, how do we interpret this in terms of top and/or side views?

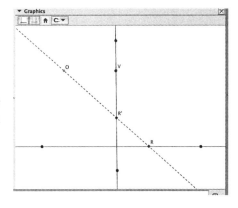

Now use the Point tool to create a point (we call it R) that we can slide along r. Use the Line tool to connect O to R, and then use the Intersection tool to create R', the image of R on b.

Click on the Move key and play around with your creation, using the following questions as a guide:

4. What happens to R' when you move O? (This is a very open-ended question; just play around and see whether you notice anything that you think is interesting.)

5. Similarly, see what (if anything) interesting happens to R' when you move r or b?

6. Can you get R' close to the point V by moving R? Why or why not? Can R' ever coincide with V?

7. If you turn one of the lines in a circle, what does the point R' do?

In the next set of steps, you will create a setup that looks something like Figure G.0. You may wish to delete points R and V if the worksheet gets too cluttered.

In your existing GEOGEBRA worksheet, create a square: under the drop-down menu of the polygon tool (which looks like a triangle), select the Regular Polygon tool. Create two points, and when GEOGEBRA asks you for the number of sides, choose "4." This gives you your square. Label the vertices of your square A, B, C, and D.

- Click on the Move tool again.
- Verify that you are able to move the square and change its size by moving the two points you used to form the square, but that clicking on the other two corners of the square does not allow you to move or change things.

Using either the Line Segment tool or the Line tool, connect the point O with each corner of the square, and use the Intersection tool to locate the image of the corners of the square on b; label these images A', B', C', and D'. (Figure G.0 uses dotted line segments between O and the square.)

Using the Parallel Line tool, locate the lines through O that are parallel to the edges of the square. (Figure G.0 uses dashed lines.) Then use the Intersection tool to locate the vanishing points for the edges of the square on b.

8. What happens to the image of the square and to its vanishing points when you move O? (This is a very open-ended question; just play around and see whether you notice anything that you think is interesting).

9. If you rotate the square in a circle (say, by clicking on one of the movable corners and moving that point in a circle about another corner), what happens to the images of the corners?

10. If you rotate the square in a circle, what happens to the lines of sight to the vanishing points? How do these lines relate to the square? How do these lines relate to each other?

11. If you rotate the square in a circle, what happens to the vanishing points themselves? Can you say anything about how these points relate to the square or to each other?

12. If one edge of the square is parallel to *b*, where are the two vanishing points?

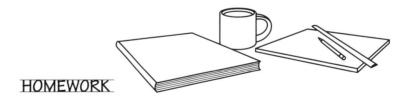

HOMEWORK

ART ASSIGNMENT

⚠ G.1. Use GEOGEBRA to create a sketch of a hallway in one-point perspective that contains at least 30 line segments depicting the following objects:
- a horizon line with a main vanishing point,
- at least two doorways,
- at least one "thing" on a wall.

All of the above objects should be done in proper one-point perspective consistent with your vanishing point and horizon line.

Your picture should look "nice." That means in part that you should hide all construction lines but leave the vanishing point visible. On the objects, hide most points for aesthetic reasons (big fat points make ugly pictures). However, you should keep visible the main vanishing point and the lines that allow you to move the horizon line.

Even more, if you move things around in your GeoGebra file ("things" being the vanishing point or corners of your doors or some such), the picture should not fall apart: the door jambs should be parallel or perpendicular to the horizon or should still head to the vanishing point.

You should submit three versions of this project:

(a) A paper version with all relevant objects (objects, horizon line, vanishing point) visible, but the construction lines hidden.

(b) Another paper version, with the objects moved around, but still all visible as before.

(c) A GeoGebra version. If your name is Sam Smith, then name your file "PG-geogebra-hallway-Sam-Smith.ggb."

Appendix R
Reference Manual

1. Window Taping: The After Math

We often analyze an image by using a *top view* and a *side view* as in Figure R.1. In this diagram, we see a *viewer*, the *picture plane* (which is drawn edge-on as a line in this diagram), and the *object*. In these two views, the object is a beach towel lying on the horizontal ground.

STATEMENTS ABOUT VANISHING POINTS OF LINES_____

- If a line is not parallel to the picture plane, then it has a vanishing point. That vanishing point is a point that is the intersection of a plane and a line: the picture plane and the line through the artist's eye that is parallel to the original line.
- A collection of lines that are parallel to each other and also parallel to the picture plane has no vanishing point(s).
- A collection of lines that are parallel to each other but not parallel to the picture plane has exactly one vanishing point.

n-POINT PERSPECTIVE _____

When we work with the perspective image of a simple, rectangular 3-D object, the definition of the "n" in "n-point perspective" is the number of the sets of edges of objects that are

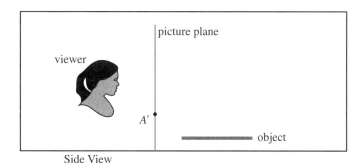

Side View

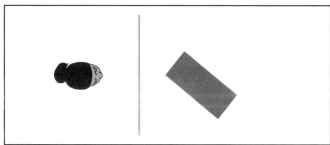

Top View

FIGURE R.1: A side view and a top view.

not parallel to the picture plane. In other words, it's the number of vanishing points of the edges. In general, we don't count vanishing points of construction lines or diagonal lines within the box when we're describing n-point perspective.

For a more complicated picture with non-rectangular objects or with objects oriented in different directions, we usually can't say the entire picture is in n-point perspective for some n, but we might be able to describe *parts* of the picture as being in n-point perspective, for varying values of n.

2. Drawing ART

Definition A *pencil* of lines in $\mathbb{R}^3$ is a collection $\mathcal{P}$ of lines satisfying either, (a) there exists a point $P \in \mathbb{R}^3$ such that $\mathcal{P} = \{p \subset \mathbb{R}^3 : P \in p\}$, or (b) there exists a line $\ell \subset \mathbb{R}^3$ such that $\mathcal{P} = \{p \subset \mathbb{R}^3 : p \parallel \ell\}$.

That is, a *pencil* of lines could be a collection of lines passing through a common point, or it could be a collection of parallel lines.

4. The Geometry of $\mathbb{R}^2$ and $\mathbb{R}^3$

Euclidean geometry

Incidence refers to meeting or touching. A point is *incident* with a line if the point is on the line. A line is *incident* with a point if it goes through the point.

Axiom 1 (Two points determine a line). *Two distinct points determine a unique line. In other words, for any two distinct points A and B, there is one and only one line incident with both A and B.*

Axiom 2 (Parallel Postulate). *Given a line and a point in a plane, the point not incident with the line, there exists a unique line in the plane that is incident with the point but not incident with the line. That is, there is a unique line in the plane that is parallel to the first line and that passes through the point.*

Definition If two or more points are incident with a common line, we say they are *collinear points*.

If two or more lines are incident with a common point, we say they are *concurrent lines*.

A *line segment* is the section of a line that is between two distinct points on the line, called *endpoints*. We denote a line segment as we would a line, except with a line segment over it, by $\overline{AB}$.

A *ray* is the section of the line that has only one endpoint and extends infinitely in one direction, denoted by $\overrightarrow{AB}$, where A is the endpoint and B is any point along the ray.

Two rays $\overrightarrow{AB}$ and $\overrightarrow{AC}$ (or line segments $\overline{AB}$ and $\overline{AC}$) that are incident at endpoint A form two sections between the rays called *angles with vertex A*, both unfortunately denoted by $\angle BAC$. (When we say "sections," we mean that if we start at ray $\overrightarrow{AB}$, we can rotate the ray about the point A in two different directions, clockwise and counterclockwise, to reach ray $\overrightarrow{AC}$. These turns sweep out two different sections of the plane. Usually, the smaller of the turns is what we mean by $\angle BAC$.)

Definition A *triangle* is a geometric figure made of three noncollinear points and the three lines determined by pairs of the points.

A *quadrangle* is a geometric figure made of four points, no three of which are collinear, and the six lines determined by pairs of the points.

Axiom 3 (Measurement). *We can choose an arbitrary line segment to have unit length equal to 1, such that every other line segment has a length relative to that unit length. Likewise, we can choose an arbitrary angle to have unit angular measurement equal to 1, such that every other angle has angular measurement relative to that unit. We will denote the length of a line segment $\overline{AB}$ by $\|AB\|$. This implies $\|AB\| = 0$ if and only if $A = B$. Although we have established the notation $\angle BAC$ for angle, we will use the same notation for the angle's measurement.*

Axiom 4 (Sum of lengths). *If A, B, and C are in that order along a line, then $\|AB\| + \|BC\| = \|AC\|$. If $\overrightarrow{AB}$, $\overrightarrow{AC}$, and $\overrightarrow{AD}$ are three concurrent rays, and if $\angle BAC$, $\angle CAD$, and $\angle BAD$ are three angles that are formed by turning in a common direction, then $\angle BAC + \angle CAD = \angle BAD$.*

Axiom 5 (Line segment copy). *If $\overline{AB}$ is a line segment and A' is a point on a line ℓ (where $\overline{AB}$ may or may not be on ℓ), then upon a given side of A' on ℓ, there is only one point B' such that $\overline{AB} \cong \overline{A'B'}$. Likewise, if $\angle BAC$ is an angle and $\overrightarrow{A'B'}$ is a ray, then upon a given side of $A'B'$, there is only one point C' such that $\angle BAC \cong \angle B'A'C'$.*

Axiom 6 (Transitivity of congruence). *If $\overline{AB} \cong \overline{A'B'}$ and $\overline{A'B'} \cong \overline{A''B''}$, then $\overline{AB} \cong \overline{A''B''}$. Likewise, if $\angle BAC \cong \angle B'A'C'$ and $\angle B'A'C' \cong \angle B''A''C''$, then $\angle BAC \cong \angle B''A''C''$.*

Definition A *transversal* of a line ℓ is a line that crosses ℓ (that is, a line not parallel to ℓ). If the transversal is not perpendicular to ℓ, notice that the angles on either side of the transversal will still sum to the measurement of two right angles. We say that a pair of angles whose measurements sum to 180 degrees are *supplementary angles*. For example, in Figure R.2, $\angle 3$ and $\angle 4$ are supplementary.

A transversal of two lines is a line that crosses both lines. In Figure R.2 we demonstrate other relationships between pairs of angles in this context:

- ∠1 and ∠3 are *corresponding angles*, as are ∠4 and ∠5,
- ∠1 and ∠2 are *vertical angles*,
- ∠2 and ∠3 are *alternate interior angles*, and
- ∠2 and ∠4 are *interior angles*.

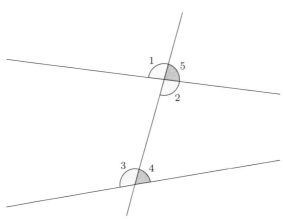

FIGURE R.2: Transversal

Euclid's Parallel Postulate: If a line segment intersects two straight lines forming two interior angles on the same side that sum to less than two right angles, then the two lines, if extended indefinitely, meet on that side on which the angles sum to less than two right angles.

Euclidean geometry revisited: Similarities and invariants

Axiom 7 (SAS). *For two triangles $\triangle ABC$ and $\triangle A'B'C'$, if two corresponding sides and the angle between them are congruent, then the triangles are congruent. In other words, if $\|AB\| = \|A'B'\|$, $\angle ABC = \angle A'B'C'$, and $\|BC\| = \|B'C'\|$, then $\triangle ABC \cong \triangle A'B'C'$.*

Axiom 8 (SSS). *For two triangles $\triangle ABC$ and $\triangle A'B'C'$, if their three corresponding sides are congruent, then the triangles are congruent.*

1. (ASA) If two corresponding angles and the side between two triangles are congruent, then the triangles are congruent.
2. (SAA) If two corresponding angles and a side not between them are congruent, then the triangles are congruent.

Axiom 9 (Similarity). *Two triangles $\triangle ABC$ and $\triangle A'B'C'$ satisfy (AAA) if and only if their corresponding sides are proportional by some linear scaling factor; that is, there exists $k > 0$ such that $\|AB\| = k \cdot \|A'B'\|$, $\|BC\| = k \cdot \|B'C'\|$, and $\|AC\| = k \cdot \|A'C'\|$.*

Ceva's Theorem. *Let $\triangle ABC$ be a triangle, and let D, E, and F be on the lines BC, CA, and AB respectively such that lines AD, BE, and FC are concurrent. Then*

$$\frac{|AF|}{|FB|} \cdot \frac{|BD|}{|DC|} \cdot \frac{|CE|}{|EA|} = \underline{\quad}.$$

Converse of Ceva's Theorem. *Let $\triangle ABC$ be a triangle, and let D, E, and F be on the lines BC, CA, and AB respectively. If*

$$\frac{|AF|}{|FB|} \cdot \frac{|BD|}{|DC|} \cdot \frac{|CE|}{|EA|} = \underline{\quad},$$

then the lines AD, BE, and FC are concurrent.

Menelaus's Theorem. *Let $\triangle ABC$ be a triangle, and let a transversal line intersect sides BC, CA, and AB at D, E, and J respectively such that D, E, and J are distinct from A, B, and C. Then*

$$\frac{|AJ|}{|JB|} \cdot \frac{|BD|}{|DC|} \cdot \frac{|CE|}{|EA|} = \underline{\quad}.$$

Definition Let $\triangle ABC$ be a triangle, and let a transversal line intersect sides BC, CA, and AB at D, E, and J respectively such that D, E, and J are distinct from A, B, and C. Denote the intersection points by $X = AD \cdot BE$ and $F = CX \cdot AB$. Then we say the points $AFBJ$ form a *harmonic set* (which we denote by H(AB, FJ)).

Harmonic Ratio Theorem. *Let A, F, B, and J be four points along a line, in that order. This set of points is a harmonic set if and only if*

$$\frac{|AF|}{|FB|} \cdot \frac{|BJ|}{|JA|} = -1.$$

5. Extended Euclidean Space: To Infinity and Beyond

Definition For any line $\ell \subset \mathbb{R}^2$, we denote the set of all lines parallel to ℓ by the symbol $[\![\ell]\!]$. We will then define a new object, denoted by $P_{[\![\ell]\!]}$, which we call the *ideal point* of ℓ.

(This notation is meant to remind us of a combination of "{}" and "||," the symbol for set and the symbol relating parallel lines: for example, "$\ell \| k$" means ℓ is parallel to k).

The *extended plane*, $\mathbb{E}^2$, consists of the points of the Euclidean plane $\mathbb{R}^2$ together with the collection of ideal points of lines in $\mathbb{R}^2$ such that the following conditions hold.

- Elements of $\mathbb{E}^2$ are *points*; a point in $\mathbb{E}^2$ is either an *ordinary* point $P \in \mathbb{R}^2$ or an *ideal* point $P_{[\![\ell]\!]}$ for some line $\ell \subset \mathbb{R}^2$.
- A line in $\mathbb{E}^2$ is either the *ideal line* ℓ_∞ (which we define to be the union of all ideal points in $\mathbb{E}^2$) or an *ordinary line* $\ell = \ell_0 \cup P_{[\![\ell_0]\!]}$ (obtained from the union of the points of a Euclidean line ℓ_0 together with the ideal point $P_{[\![\ell_0]\!]}$ of that line).

In the same way that we defined the extended plane as an extended Euclidean plane, we may define *extended Euclidean space*, $\mathbb{E}^3$, as an extension of $\mathbb{R}^3$, as in the definitions below.

Definition For any plane $\alpha \subset \mathbb{R}^3$, we denote the set of all planes parallel to α by the symbol $[\![\alpha]\!]$. We then define a new object, denoted by $\ell_{[\![\alpha]\!]}$, which we call the *ideal line* of α. This ideal line contains a collection of ideal points: we say $P_{[\![k]\!]} \in \ell_{[\![\alpha]\!]}$ whenever the Euclidean line k is parallel to the plane α.

Definition. *Extended space*, $\mathbb{E}^3$, consists of the points of Euclidean space $\mathbb{R}^3$ together with the collection of all ideal points of lines in $\mathbb{R}^3$ such that the following conditions hold.

- Elements of $\mathbb{E}^3$ are *points*; a point in $\mathbb{E}^3$ is either a point in $P \in \mathbb{R}^3$ (called an *ordinary point*) or an ideal point $P_{[\![\ell]\!]}$ for some line $\ell \subset \mathbb{R}^3$.
- A line in $\mathbb{E}^3$ is either an *ideal line* $\ell_{[\![\alpha]\!]}$ for some plane $\alpha \subset \mathbb{R}^3$ or an *ordinary line* $\ell = \ell_0 \cup P_{[\![\ell_0]\!]}$ (obtained from the union of the points of a Euclidean line $\ell_0 \subset \mathbb{R}^3$ together with the ideal point $P_{[\![\ell_0]\!]}$ of that line).
- A plane in $\mathbb{E}^3$ is either the *ideal plane* α_∞ (which we define to be the union of all ideal points in $\mathbb{E}^3$) or an *ordinary plane* $\alpha = \alpha_0 \cup \ell_{[\![\alpha_0]\!]}$ (obtained from the union of the points of a Euclidean plane $\alpha_0 \subset \mathbb{R}^3$ together with the points in the ideal line $\ell_{[\![\alpha_0]\!]}$ of that plane).

LEMMAS ABOUT $\mathbb{E}^2$

1. Every pair of distinct lines in $\mathbb{E}^2$ intersects in one and only one point.
2. Every pair of distinct points in $\mathbb{E}^2$ determines exactly one line.

THEOREMS ABOUT $\mathbb{E}^3$

Theorem E1: Two distinct points in $\mathbb{E}^3$ determine a unique line.
Theorem E2: Two distinct lines in $\mathbb{E}^3$ lie in the same plane if and only if the lines intersect in exactly one point.
Theorem E3: Two distinct planes in $\mathbb{E}^3$ determine a unique line.
Theorem E4: A plane and a line not on the plane determine a unique point.
Theorem E5: Three noncollinear points in $\mathbb{E}^3$ determine exactly one plane.

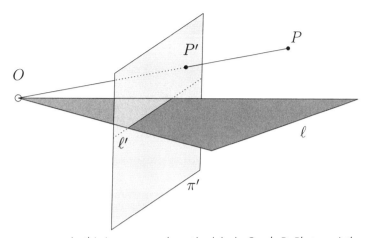

FIGURE R.3: In this image, we show the labels O, π', P, P', ℓ, and ℓ' that illustrate part of a mesh map.

6. Of Meshes and Maps

RELATING MESH MAPS, EXTENDED REAL SPACE, AND VANISHING POINTS

A vanishing point is an ordinary point, not an ideal point. But a vanishing point is the *image* of an ideal point under a mesh map.

Definition A *mesh* $\mathcal{M} = \{P_1, \ldots, P_k, \ell_1, \ldots, \ell_n\}$ is a collection of points and lines in $\mathbb{E}^3$ such that every point in $\mathcal{M}$ is incident with at least two lines in $\mathcal{M}$ and every line in $\mathcal{M}$ is incident with at least two points in $\mathcal{M}$.

Definition Given a mesh $\mathcal{M}$ in $\mathbb{E}^3$, a plane $\omega' \subset \mathbb{E}^3$, and a point $O \in \mathbb{E}^3$ not incident with $\mathcal{M}$ or ω', we can define a *mesh map* as

$$\prime = \mathcal{M} \to \mathcal{M}' \subset \omega'$$

with center O, taking points to points and lines to lines such that

- O, P, P' are collinear for every point $P \in \mathcal{M}$, and
- O, ℓ, ℓ' are coplanar for every line $\ell \in \mathcal{M}$.

LEMMAS ABOUT MESH MAPS

In Lemmas 13–15, we assume we have a mesh map $\prime = \mathcal{M} \to \mathcal{M}'$ as described in the definition.

13. If $Q \in \ell$ and if both Q and ℓ are in $\mathcal{M}$, then $Q' \in \ell'$.

14. If P and Q are distinct points in $\mathcal{M}$ with PQ also in $\mathcal{M}$, then $P'Q' = (PQ)'$. That is, the image of the line between two points is the line between the two images of the points.

15. If k and ℓ are distinct lines in $\mathcal{M}$ with $k \cdot \ell$ also in $\mathcal{M}$, then $k' \cdot \ell' \subseteq (k \cdot \ell)'$. That is, the image of the intersection of two lines lies in the intersection of the images of the lines.

Definition A *triangle* is a mesh with exactly three points (equivalently, with exactly three lines).

A *quadrangle* is a set of four coplanar points, such that no three are collinear, as well as the ____ lines defined by pairs of these four points.

A *complete quadrangle* is a quadrangle, plus the other three points of intersection defined by the ____ lines. These three points are called the *diagonal points*.

---

7. Desargues's Theorem

Definition We say a point and line are *incident* if the point lies on the line.

Definition Suppose $\mathcal{M}_1$ and $\mathcal{M}_2$ are two collections of points and lines in $\mathbb{E}^3$, each of them planar. We say $\mathcal{M}_1$ and $\mathcal{M}_2$ are *perspective from a point* if there is a point O (called the *center*) and a one-to-one correspondence between the points of $\mathcal{M}_1$ and $\mathcal{M}_2$ such that each of the lines through the corresponding points is incident with the point O.

Definition Suppose $\mathcal{M}_1$ and $\mathcal{M}_2$ are two collections of points and lines in $\mathbb{E}^3$, each of them planar. We say $\mathcal{M}_1$ and $\mathcal{M}_2$ are *perspective from a line* if there is a line o (called the *axis*) and a one-to-one correspondence between the lines of $\mathcal{M}_1$ and $\mathcal{M}_2$ so that each of the intersections of corresponding lines is incident with the line o.

Desargues's Theorem. *If two triangles in $\mathbb{E}^3$ are perspective from a point, then they are perspective from a line.*

Desargues's Theorem and Its Converse. *Two triangles in $\mathbb{E}^3$ are perspective from a line if and only if they are perspective from a point.*

Perspective Proposition. Suppose two meshes $\mathcal{M}$ and $\mathcal{N}$ in $\mathbb{E}^3$ are perspective from some point P. Then we can create a new, larger mesh

$$\mathcal{P} = \mathcal{M} \cup \mathcal{N} \cup \{P\} \cup \{\text{lines of perspectivity}\}$$

by taking the union of our original meshes together with the point P and the lines that connect P to the points of $\mathcal{M}$ and $\mathcal{N}$. Choose some point O not incident with $\mathcal{P}$ and some plane ω' not incident with O. This gives us a mesh map $\prime : \mathcal{P} \to \omega'$. Then the images $\mathcal{M}'$ and $\mathcal{N}'$ are perspective from a point in the plane ω'.

8. Collineations

Definition A *perspective collineation* is a function $h: \omega \to \omega'$ between planes $\omega, \omega' \subset \mathbb{E}^3$ (possibly the same plane) that takes points to points and lines to lines, satisfying both of these conditions:

- there exists a point $O \in \mathbb{E}^3$ (called the *center*) such that O, P, and $h(P)$ are collinear for every point $P \in \omega$, and
- there exists a line $o \subset \omega \cdot \omega'$ (called the *axis*) such that o, ℓ, and $h(\ell)$ are coincident for every line $\ell \subset \omega$.

If $h(O) \in o$, then we say h is an *elation*; otherwise h is a *homology*.

We say the collineation *fixes* a point P or a line ℓ (or that P or ℓ is *fixed*) if $h(P) = P$ or $h(\ell) = \ell$.

Three Points Theorem: *A perspective collineation is determined by three points (non-collinear) and their images.*

Center-Axis-Point Theorem: *A perspective collineation is determined by its center, its axis, and a point not on the axis together with its image.*

REFLECTING ON HOMOLOGIES AND HARMONIC SETS

Definition A *period-2 homology* is a homology h for which $h(h(X)) = X$ for every $X \in \mathbb{E}^2$.

A set of four distinct collinear points A, B, C, D is called a *harmonic set*—which we write as $H(AC, BD)$—if there exists some quadrangle for which one pair of opposite lines is coincident at A, another pair of opposite lines is coincident at C, and the remaining two lines contain B and D respectively.

Suppose h is a homology with center O and axis o. For every $X \in \mathbb{E}^2 \setminus \{O\}$, we denote by X_o the intersection of the lines o and OX. We say h is a *harmonic homology* if the points $X, X_o, h(X), O$ form a harmonic set: that is, if $H(Xh(X), X_o O)$.

Theorem. *Any period-2 homology is fully determined by two points and their images, provided no three of those four points are collinear.*

Theorem. *Every period-2 homology is a harmonic homology.*

Elations (or, How to be transported in a math class)

Recall that we say that a function e from $\mathbb{E}^2$ back to itself is an *elation* if, for every mesh $\mathcal{M}$ in $\mathbb{E}^2$, we have $\mathcal{M}$ and its image $e(\mathcal{M})$ are perspective from both a point (center) and a line (axis), and if the center lies on the axis.

Elation Theorem. *An elation is completely determined by its axis, its center, and by one pair of corresponding points.*

9. Dynamic Cubes and Viewing Distance

> **Definition** For a perspective picture, the *viewing target T* is the point in the picture plane ω that is closest to the center O.
> That is, OT is perpendicular to ω. The *viewing distance* $|OT|$ is the distance from the center to the picture plane.

To determine the viewing distance for a one-point picture, we can use the image of a horizontal square. Using that square, locate the primary vanishing point V, which in turn gives us the horizon line h. From there, we can locate the vanishing point W of a diagonal of the square; the viewing distance is $|VW|$. See Figures R.4 and R.5.

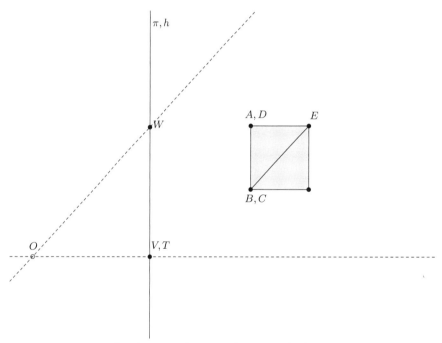

FIGURE R.4: Top view for drawing a horizontal square. The line OV is parallel to one edge of the square; the line OW is parallel to the diagonal on the top of the square. Therefore the triangle OTW is an isosceles triangle; that is, the viewing distance $|OT|$ is equal to $|TW| = |VW|$.

10. Drawing Boxes and Cubes in Two-Point Perspective

To determine the viewing target and distance for a picture in two-point perspective, we can use the image of a rectangle. The viewer stands on the semicircle whose diameter is the line segment between the vanishing points of the edges of that rectangle. If the rectangle is acutally a square, then the viewer also stands on the semi-circle whose diameter is the line segment between the vanishing points of the diagonals of that square. See Figure R.6.

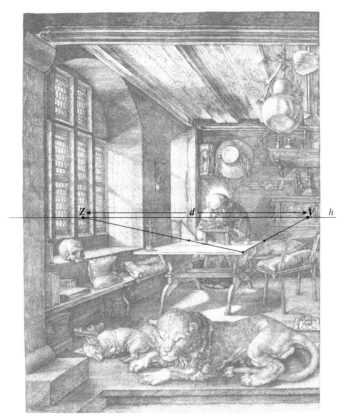

FIGURE R.5: The viewing distance for *St. Jerome* is fairly small; the diagonal vanishing point (assuming a square table top) is in the windows.

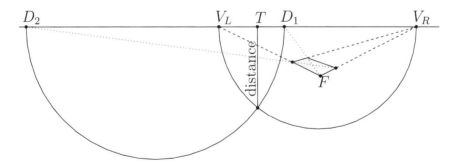

FIGURE R.6: The viewing target *T* and the line segment representing viewing distance.

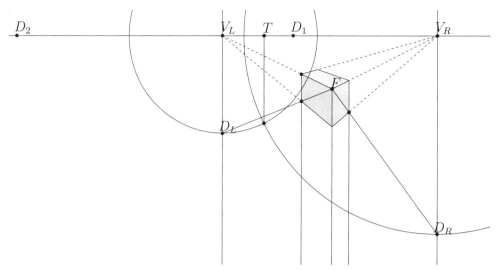

FIGURE R.7: Constructing the two vertical faces of a cube.

11. Perspective By the Numbers

DISCOVERING THE CROSS RATIO

Definition Given any two ordinary points $A, B \in \mathbb{E}^3$, we denote the distance between these points by the symbol $\|AB\|$. We will use the symbol $|AB|$ to denote the *directed distance* between these points. That is, we assign (arbitrarily) a direction to the line containing the points; we say $|AB| = \|AB\| > 0$ if B comes after A, and $|AB| = -\|AB\| < 0$ if B comes before A. Note that $|AB|/|BA| = -1$ always.

Given four ordinary collinear points A, B, C, D, we define the *cross ratio* (AC, BD) by

$$(AC, BD) = \frac{|AB|}{|BC|} \cdot \frac{|CD|}{|DA|}.$$

Definition When collinear points A, B, C, D satisfy $(AC, BD) = -1$, we say these points form a *harmonic set*; equivalently, we might write $\mathrm{H}(AC, BD)$.

EVES'S THEOREM

Definition The ratio of a product of directed segments to another product of directed segments, where all the segments lie in one plane, is called an *h-expression* if it has the following properties:

1. If we replace $|AB|$ (or $\|AB\|$, if we use all positive values) by $A \times B$, and similarly treat all other segments appearing in the ratio, and then regard

the resulting expression as an algebraic one in the letters *A*, *B*, etc., these letters can all be cancelled out.

2. If we replace |*AB*| (or ‖*AB*‖, if we use all positive values) by a letter, say ℓ, representing the line *AB*, and similarly treat all other segments appearing in the ratio, and then regard the resulting expression as an algebraic one in the letters ℓ, etc., these letters can all be cancelled out.

Eves's Theorem: *The value of an h-expression is invariant under any projective transformation.*

AN ANGLE ON PERSPECTIVE: CASEY'S THEOREM

Definition Given four distinct collinear points, denoted in order by *A*, *B*, *C*, and *D*. Let *O* be an intersection point of the circles with diameters *AC* and *BD*. Then the angle ∠*BOC* is the *Casey angle* of those four points.

Casey's Theorem: *The Casey angle is a projective invariant.*

12. Coordinate Geometry

Definition We define a *homogeneous point* $(x:y:z)$, or equivalently a *point in the projective plane*, to be the equivalence class of points in $\mathbb{R}^3$ minus the origin under the equivalence $\sim$, so that

$$(cx:cy:cz) = (x:y:z)$$

for $c \neq 0$.

Note that we use slightly different notation for two different kinds of objects. For the real point $(x, y, z) \in \mathbb{R}^3$, we write $(cx, cy, cz) \sim (x, y, z)$; for the homogeneous point $(x:y:z)$ in the projective plane, we write $(cx:cy:cz) = (x:y:z)$.

A point in the projective plane represents a line through the origin in real space. By analogy, a line in the projective plane represents a plane through the origin in real space.

Definition We define a *homogeneous line* $[a:b:c]$, equivalently a *line in the projective plane*, to be the equivalence class of lines that lie in a real plane with equation $ax + by + cz = 0$. It follows that $[a:b:c] = [wa:wb:wc]$ for any nonzero $w \in \mathbb{R}$.

Appendix W
Writing Mathematical Prose

W.1 Getting Started

W.1.1 *WHY* WE WRITE PROOFS, AND WHAT THAT MEANS FOR *HOW* YOU WRITE PROOFS

A proof is a communication between you and your reader. You are trying to convince another person that a certain mathematical statement is true (or, perhaps, that it is false).

That goal of communication has several consequences for how you write. In particular:

- You should aim your explanations so that they are readable and understandable by other people in the class; try to read your papers through a student's eyes.
- Professional mathematicians often get help and ideas from other places; then they give acknowledgment where due. When you get help from another student—either an idea of how to approach an argument or help proofreading—you should give acknowledgement of this help. Although you should try to solve the homework without resorting to looking up the answers elsewhere, if you discover interesting related material in textbooks or on the web, you should provide a proper citation. Thanks and citations come at the end of your homework, before the endnotes.
- You should feel free to ask your professor questions. You should add those as footnotes or endnotes. If instructors know what questions you have, they can give you better feedback.
- You should leave space for comments. In particular, you should leave 1" margins on all sides of the paper. For similar reasons, leave small amounts of space between lines or paragraphs.

Your mathematical proofs should look "good" in the sense of being easy for other people to read. Good handwriting and/or a nice font matter.

W.1.2 MECHANICS AND CONVENTIONS

Start with a header that gives basic "bookkeeping" information: your full name(s), the date, and the title of the assignment.

In the same way that a good essay begins with a thesis statement, a mathematical proof should begin with the theorem you are proving or the statement you are disproving. In formal mathematics, we place the word "Theorem" in bold face, followed by a period, and italicize the statement of the theorem itself.

> **Theorem.** *If lines ℓ and k are parallel to each other but not parallel to the picture plane, then their images have the same vanishing point.*

or

> **Claim (false).** *If two distinct lines in $\mathbf{R}^3$ do not intersect, then they are parallel.*

After the statement, begin the proof or counterexample on a new line with the appropriate title. The word "Proof" is italicized, but the proof itself is not.

Proof. Suppose ℓ and k are parallel to the plane ω' …

A proof ends with a black box, justified right, as here.　　　　　　　　　■

YOUR TURN: ANSWER THESE QUESTIONS ABOUT HOMEWORK FORMAT.

_____

1. What goes here?
2. What goes here?
3. What goes here?

4. **What word (or pair of words)**[1] **appears in bold?** 5. What punctuation comes right after this?

6. *What word*[2] *appears in italics?* 7. What punctuation appears after that word?

8. Don't underline words when you use the computer—professionals frown upon that! Underlining text is the old-fashioned way that authors who hand-wrote or typed their manuscripts let professional typesetters know which words to emphasize.

If you write your homework by hand, does an underlined word properly indicate italics or bold?

9. We all want to save trees and use as little paper as possible, so it's tempting to use all available space on the paper. In spite of that temptation, why is it a good idea to leave a bit of space between lines and between paragraphs?

10. Name three possible reasons that it's a good idea to have a classmate proofread your work before you turn it in.

11. What symbol comes at the end of your proof?

12. What important professional information goes here, at the end of your write-up?
13. What important communication goes in the footnotes?

[1] There are two possibilities.
[2] Again, there are two possibilities.

W.2 Pronouns and Active Voice

THE PERILS OF "IT," "THIS," AND "IS"

Pronouns and "to be" verbs make your exposition ambiguous and wordy. Consider the following nine-word sentence:

```
It is easy for us to solve the equation.
```

What does the word "it" refer to? *Nothing*; and therefore beginning the sentence with "It" creates a possible source of confusion for the reader. A somewhat better version, only seven words long, replaces "it" with a noun phrase:

```
Solving the equation is easy for us.
```

But even better would be to replace that neutral "is" with the action of "solving":

```
We can easily solve the equation.                        ☺
```

1. Rewrite the following sentence to remove "it" and "is":

```
It is clear that it is impossible to find a real root of
this quadratic; it is always above the x-axis.           ☹
```
                                                                                              ☺

Mathematicians prefer to present our arguments in the active voice, in the present tense. We use "we" even when there is only one author: the pronoun means "the readers and I."

2. Rewrite the following sentence to match the preferred mathematical style:

```
Both sides of the equation were divided by 2 to get x = 5.  ☹
```
                                                                                              ☺

---

3. Which overused pronoun makes the following paragraph confusing to read?

```
The plan view shows the viewer, the picture plane, and a
wall. This is a vertical plane, so it is parallel to the
wall. This means that a line on the wall is parallel to its
image on the picture plane, and so this will give the artist
a picture that is similar to the one on the wall. This is why
the two pictures look the same.
```

4. Rewrite the above paragraph to include no instances of the word "this."

Are "it," "this," and "is" always bad? Should we avoid every instance of these words?
No. We can safely use "it" or "this" when

- we are not overusing these words (there are no other instances of these words nearby), and
- the noun immediately preceding the pronoun means the same thing:

To change the color of the <u>room</u>, we paint <u>it</u>.   ☺

With "this," sometimes we can help our reader by adding a noun:

<u>This</u> is why the two pictures look the same.   ☹

<u>This argument</u> explains why the two pictures look the same.   ☺

Still, too many instances of these words might indicate you haven't thought as hard as you ought to about presenting a clear argument. if your writing has many instances of "it" and "this," you should to revisit your proof to see whether you can make your language and exposition more specific and clear.

W.3 Introducing and Using Variables, Constants, and Other Mathematical Symbols

Introduce each variable or constant the first time you use it. Good words to introduce symbols include "let," "choose," "call," and "denote." Here are some examples:

Suppose O is the center of the projection ... Let X and Y be the corners of the first door ... Then X and Y must lie on a common line; call that line ℓ. ☺

If you use letters or symbols to name your objects, you need to introduce the names in writing even if you have a diagram that includes those names. We italicize variables to tell them apart from regular letters.

1. Prove that the sum of three even numbers is an even number. In this proof, introduce any variables or constants you might need.

If you use a quantity only a few times, see if you can get away with not assigning it a variable.

2. Rewrite the following sentence without using variables (but you can still use numbers). You should be able to do this in a way that shortens the sentence considerably without losing mathematical content.

We see that $A = (1/2) \times h \times b$, where A stands for the area of the triangle, b stands for the base of the triangle, and h stands for the height of the triangle, and so $A = (1/2) \times 3 \times 4 = 6$ square inches. ☹

If you don't use LaTeX but wish to write your homework on the computer, *Microsoft Word* has an equation editor: pull down the **Insert** menu, select **Object**, and then **Equation**. If you don't have an equation editor, you may try using tabs and fancy fooling, or you may wish to write the mathematics in by hand.

Using symbols in prose

Do not confuse mathematical symbols for words ("=," "≠," and " $\Longrightarrow$ " are especially common examples of symbols misused as words). Use equal signs when you state formulas or equations, but not as the verbs of your sentence.

3. What symbol/word confusions appear in the following two sentences?

   > We let V = volume of a single mug and n = the # of mugs.
   > Then the formula for R, the total amount of root beer we can
   > pour, is R is nV. ✕

4. Rewrite the sentences from question 3 with appropriate use of symbols and words.

When you write at the chalkboard or whiteboard, writing "$A \subset B \Longrightarrow A \cap B \neq \emptyset$," is appropriate. But on paper, you should spell out "implies": "That $A \subset B$ implies $A \cap B \neq \emptyset$", or "Because $A \subset B$, we get $A \cap B \neq \emptyset$." Similarly, in written work we do not abbreviate common mathematical phrases such as "wlog" (without loss of generality).

5. Shorten this sentence (using mathematical symbols and common abbreviations) as you might write it on a chalkboard.

   > For all $\epsilon > 0$, there is some $\delta > 0$ such that $|f(x) - L| < \epsilon$
   > whenever $|x - a| < \delta$.

6. Rewrite the following in a way that is suitable for a written document.

   > WLOG, assume $L \geq 0$. Then $\exists N \in \mathbb{Z}$ s.t. $n > N \Longrightarrow L - a_n > 0$.

7. Rewrite these sentences so that they do not begin with a mathematical symbol:

   > H is the height of the triangle and b is the base, so
   > $A = \frac{1}{2} \cdot h \cdot b$.

   > x stands for the length of the gusset, measured in inches.

W.4 Punctuation with Algebraic Expressions in the Sentence

1. What is wrong with the punctuation in the following sentences? Fix the punctuation.
   (a) My favorite pet is: my dog
   (b) I ate the following breakfast foods. Waffles and pancakes.
   (c) We have three cats

Mathematical formulas are like clauses or sentences: they need proper punctuation. Put periods at the end of a computation if the computation ends the sentence; use commas if it doesn't.

Here is a little-known colon fact: a colon properly comes after a noun that ends a clause and that describes another noun (or list of nouns). It does not come after a verb.

We have five animals: a cow, a donkey, and three chickens. (colon after "animals")
We have two sons and four daughters.   (no colon)
I gave the book to my friend Mary.   (no colon)

2. Which of these sentences require a colon?
   (a) To determine how much paint to buy, we use the following formula

   $$A = LH.$$

   (b) To determine how much paint to buy, we use the formula

   $$A = LH.$$

   (c) To determine how much paint to buy, we use the formula for the area of a rectangle

   $$A = LH.$$

   (d) To determine how much paint to buy, we use

   $$A = LH.$$

3. What is wrong with the punctuation in the following sentences? Fix the punctuation.
   (a) The formula for volume is:

   $$V = LWH$$

   (b) I then plugged in $t = 3$ and found the formula for velocity.

   $$v = g/2 \cdot 3^2.$$

   (c) Einstein developed the famous formula

   $$E = MC^2$$

   where E stands for energy, M=mass, and C is the speed of light.

Hyphens

Use a hyphen with a compound adjective that precedes the noun it modifies. That is, we would write,

```
They placed the three-cornered hat in the box.                    ✓
```

It was not a "three hat," nor a "cornered hat," so we use the hyphen to make it a "three-cornered hat." But we do *not* use a hyphen when the compound adjective stands apart from the noun in the sentence:

```
They only wear hats that are three cornered.                      ✓
```

If instead the adjectives are serial (that is, if we could place "and" between them), we use commas:

```
Create your drawing on a large, flat surface.                     ✓
```

Which of the following sentences need hyphens and/or commas between the underlined adjectives?

1. The artist drew the picture in one point perspective.
2. The figure seemed real even though it was two dimensional.
3. A high definition image takes a lot of memory, but it reproduces well.
4. The well beloved mathematician was well prepared to give a moving speech.
5. Clifford was a big red dog owned by a big hearted family who were open minded about size.

W.5   Paragraphs and Lines

1. Each of the following examples of a mathematical argument contains exactly the same sentences as the other examples. The only difference between the examples is the way the sentences are grouped (or not) into paragraphs. Which example has the best grouping, and why?

Example 1.

We believe that objects that are further away from us appear to be smaller.

When we observe same-sized objects with just one eye, then each object has just one image on the canvas.

The size of each image depends on the distance of the object from the canvas and the eye.

Therefore, the further object will appear smaller.

On the other hand, Leonardo da Vinci in his *Codex Urbinas* described why this is not always true when we observe the world with both eyes open.

If we observe two objects, both smaller than the distance between our two eyes, then each object has two images.

The images of the closer object will be close together.

The images of the far object are further apart than the images of the close object, taking up more space on the canvas.

Therefore, the image of the far object will appear larger.

Example 2.

We believe that objects that are further away from us appear to be smaller. When we observe same-sized objects with just one eye, then each object has just one image on the canvas. The size of each image depends on the distance of the object from the canvas and the eye. Therefore, the further object will appear smaller.

On the other hand, Leonardo da Vinci in his *Codex Urbinas* described why this is not always true when we observe the world with both eyes open. If we observe two objects, both smaller than the distance between our two eyes, then each object has two images. The images of the closer object will be close together. The images of the far object are further apart than the images of the close object, taking up more space on the canvas. Therefore, the image of the far object will appear larger.

Example 3.

We believe that objects that are further away from us appear to be smaller. When we observe same-sized objects with just one eye, then each object has just one image on the canvas. The size of each image depends on the distance of the object from the canvas and the eye. Therefore, the further object will appear smaller. On the other hand, Leonardo da Vinci in his *Codex Urbinas* described why this is not always true when we observe the world with both eyes open. If we observe two objects, both smaller than the distance between our two eyes, then each object has two images. The images of the closer object will be close together. The images of the far object are further apart than the images of the close object, taking up more space on the canvas. Therefore, the image of the far object will appear larger.

First and last sentences

In the same way that the points of emphasis in a sentence lie near first and last words (good writers do not "bury" important words in the middle of the sentence), the first and last sentences are natural places of emphasis in a paragraph. The first sentence of a paragraph sometimes contains a transitional phrase ("On the other hand," "In the same way that we located the vanishing points for the first box," "The next step is ...") and then explains the topic of the paragraph. In the middle of the paragraph come the supporting details. The last sentence concludes and summarizes the idea of the paragraph clearly. In this way, a well-designed paragraph structure helps your reader move easily from one logical idea in your argument to the next.

    2. Put these sentences in order to form a paragraph.
      (a) The moon lies 238,900 miles from earth, and it has a diameter of 2,159 miles.
      (b) Solving for m, we see that the artist should draw the moon about one third of an inch across, much smaller than most people would expect.
      (c) By using similar triangles, we see

$$\frac{m}{36} = \frac{2159}{238,900} \approx 0.009037.$$

      (d) In the plan view, we see that the artist is 1 yard (36 inches) from the canvas; let m stand for the size of the image of the moon.
      (e) Based on the plan view we described above, we can find a formula for the size of the image of the moon.

Centered equations

Another way to use line breaks to help readability is to add space around important mathematical formulas. Putting important or complicated formulas on a line of their own (centered) makes them much easier to read; simple and short formulas can appear in line with the rest of the text.

Hard to read:
The number of bricks in a pyramid with n layers is $1+2+\cdots+n= n(n+1)/2$. Therefore, a pyramid with 100 layers contains $100(101)/2= 5,050$ bricks. ☹

Easier to read:
The number of bricks in a pyramid with n layers is

$$1+2+\cdots+n=n(n+1)/2.$$

Therefore, a pyramid with 100 layers contains $100(101)/2=5,050$ bricks. ☺

W.6 Figures

We will use many diagrams and figures in proofs in this class. A figure is not the same as a proof, but a well-drawn, well-labeled figure can be a huge help for the reader. In particular, a good figure can illustrate the text and make your written argument easier to follow.

Here are some common mistakes students make when including figures.

× *Making the diagram overly intricate or too small to read.* Legibility matters. A multistep proof might include several figures, each illustrating one aspect of the argument.

× *Writing mathematical formulas or performing computations as part of the drawing.* Remember that your figure should illustrate your written argument, not replace it.

× *Adding inconsistent elements.* For example, if the text describing volume and temperature uses variables V and T, a graph that illustrates their relationship should not label the axes with the variables x and y.

× *Squeezing the actual information in a small area, while leaving large blank areas in the figure.* For example, if you draw a graph of $y = \sin(x)$, the y-axis should not run from -10 to 10.

W.6.1 FORMATTING THE FIGURE

Center each figure horizontally on the page. Do not run into the margins. Under each figure, add a caption, also centered horizontally. The caption begins with the word "Figure," a number, and a colon. After the colon, write a helpful description of the picture followed by a period.

Leave space between the caption and whatever follows (the next figure or the text).

$$\&\&\&\&\&\&\&\&\&\&$$
$$\&\&\&\&\&\&\&\&\&\&$$
$$\&\&\&\&\&\&\&\&\&\&$$
$$\&\&\&\&\&\&\&\&\&\&$$
$$\&\&\&\&\&\&\&\&\&\&$$

FIGURE W.1: Drawing a box with ampersands is easy if you cut and paste.

W.6.2 REFERRING TO FIGURES

A figure is not part of a sentence, so we do not "end" the sentence with a figure by writing,

```
Such and such is true, as we see in the following figure:        ×
```

Instead, write the name of the figure and use a period:

```
Such and such is true, as we see in Figure 3.                    ✓
```

Notice the capital letter in "Figure." Do not abbreviate the word "Figure."

1. Find at least six ways that the formatting or presentation of the diagrams below might be improved.

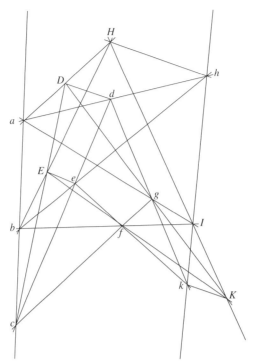

FIGURE 1: Bosse's diagram shows that two triangles perspective from a point are perspective from a line.

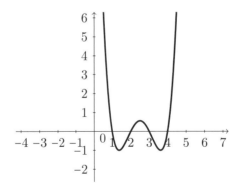

FIGURE 2: A graph of $y = f(x)$

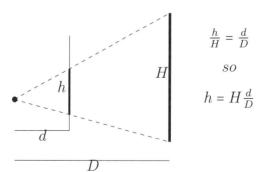

$$\frac{h}{H} = \frac{d}{D}$$

so

$$h = H\frac{d}{D}$$

FIGURE 3: Developing a formula for the size h of an image.

Acknowledgments

Putting this book together has been a labor of love, and the authors are incredibly grateful to many people who labored over the material and who loved projective geometry with us along the way. First and foremost among these are our many students, who field-tested this material, who gave us feedback directly and indirectly, and who buoyed our hopes that combining perspective art with projective geometry could be a rigorous, enjoyable way to encounter deep new mathematics.

We would especially like to thank a group of students whose work with us went above and beyond. In 2012, just as we were getting started on this project, Stephanie Douglas read Coxeter's *Projective Geometry* as an Independent Study project with Crannell; her insights led to the shadow interpretation of Desargues's theorem that we use in Chapter 7. Crannell deepened her understanding of the material further in subsequent independent studies with her students Anton Arapin, Ellie Burkett, Mariah Eble Yike Gong, Julie Heyman, Arielle Leider, Weihan Lu, Elana Machlis, Andrew Miller, Liz Sullivan John Schrieber, and Yanlin Yang. We doff our caps to Greg Bolet, who pointed out what should have been obvious (that we need to leave space after questions for people to write their answers).

In 2008, 2012, 2014, and 2017, Futamura had the pleasure of exploring and teaching connections between projective geometry and perspective drawing with her students, who helped provide valuable insight and feedback. In particular, Stephen Foster and Tommy Rogers in the first summer class in 2008 helped brainstorm early connections. Robert Lehr in 2012 not only provided valuable feedback for the textbook but also helped find a new use for the cross ratio. Kelsey Welden, Heather Gronewald, and Nathaniel Leeds in 2012 and 2014 gave insight into how this material could be taught in the high school setting. Sam Vardy provided a number of useful suggestions in 2018 for improving the nearly finalized version of the textbook, along with an interesting connection to graph theory. A number of their ideas appear in the instructor's manual as examples of student projects. Other students who enthusiastically tested the book, provided helpful feedback, developed useful GEOGEBRA constructions, and inspired us with creative projects include Katie Dyo, Hannah Freeman, Ryan Gallo, Ryan Galloway, Megan Grimes, Bonnie Henderson, Christi Ho, Isaac Hopkins, Stan Kannegieter, Yasmin Leon, Megan Myers, Ryan Ogden, Whitney McColly, Will Nguyen, Penny Phan, Will Price, Jeremy Snyder, Aiden Steinle, Julia Sykora, and Perrin Turk.

Beyond offering advice, insight, and wisdom, several brave students graciously agreed to share their artwork. We thank the following students for their permission to use their work in this book and in the instructor's manual:

- Figure 1.0 shows, with his permission, Yuxun Sun taping a window;
- Figure 1.4: with permission of Anthony Nocket and Kevin Toenboehm;
- Figure 6.4: with permission of Biyang Sun;
- Figure 2.9 (Instructor's Manual): with permission of Gabrielle Connor;
- Figure 3.3 (Instructor's Manual): with permission of Giorgi Agulashvili;
- Figure 5.4 (Instructor's Manual): with permission of Marshall Dugan and Huiyan Anna Xiong.

The Mathematical Association of America (MAA) supported this work directly and indirectly in numerous ways. It was at an MAA minicourse in 2008 that Marc and Annalisa met Fumiko, launching a mutually rewarding collaboration. The MAA has since provided invaluable administrative support by encouraging our subsequent minicourses, by accepting and inviting us to lecture at national meetings, and by supporting SIGMAA Arts. In these and many other ways, the MAA has allowed us to engage with this material with many other mathematicians. More particularly, the MAA has granted us permission to reuse images (Figures 11.16–11.18) from a paper on projective numerical invariants that Marc had published in its *Mathematics Magazine*.

We are indebted to the many faculty members who have encountered previous versions of this material—online, through our talks, or in our minicourses—and have given us pointed and useful feedback. Among these, Ray Rosentrater of Westmont College deserves plaudits and bags of chocolate; he field-tested our materials with his students (without our even asking him to!), and gave us careful, insightful, and useful suggestions for improvement. Andrius Tamulis of Governors State University offered us the photo of a truck delivering a load of Trompe L'Oeil (Figure 9.2).

The referees who read earlier drafts of this manuscript were encouraging and brimming with good advice. One of the (many) useful anomalies these referees brought to our attention was spelling inconsistencies. In that vein, we doff our bonnets to Anil Venkatesh for pointing out the reason for the mysterious (to us) double "l" in "collineation," and to Ryan Fowler for elaborating on the etymology of the word.

Vickie Kearn, our editor at Princeton University Press, has supported, mentored, encouraged, and fed us through two different book projects now; we couldn't wish for a better editor, but nonetheless we wish her all the best in her retirement. Lauren Bucca, also of the Press, answered our many picky and clueless questions with grace and precision.

This work was supported by NSF TUES Grant DUE-1140135.

Bibliography

[1] Camera Obscura. *The Universal Magazine* (a journal published in the early 1880s featuring "Original Communications in History, Philosophy, the Belles Lettres, Politics, Amusements, &c. & c."), from *Optics: the principle of the camera obscura. Engraving.* Wellcome Collection, Creative Commons Attribution (CC BY 4.0) terms and conditions https://creativecommons .org/licenses/by/4.0.

[2] 9GAG. Cat on Stairs. The original cat-on-stairs photo seems to have appeared at the site http://9gag.com/gag/azEP79x?ref=fbp in the spring of 2015. Many other sources reproduced it, including the following:

- *National Geographic*, http://news.nationalgeographic.com/2015/04 /150409-step-cat-up-down-staircase-viral-photo-science/
- *Business Insider*, http://www.businessinsider.com/ cat-on-stairs-optical-illusion-2015-4
- CNet (with a real professor!!!! to explain the illusion), http://www.cnet.com/news /professor-explains-whether-cat-is-going-up-or-down-stairs/
- ABC News, http://abcnews.go.com/Lifestyle/photo-cat-stairs-confuses-internet /story?id=30225475 and
- *USA Today*, https://www.usatoday.com/story/news/nation/2015/04/10 /cat-going-up-or-down-steps/25568635/

[3] Leon Battista Alberti. *On Painting.* (J. R. Spencer, translator). Yale University Press, New Haven, CT, 1966. http://www.noteaccess.com/Texts/Alberti/.

[4] Kirsti Andersen. *The Geometry of Art: The History of the Mathematical Theory of Perspective from Alberti to Monge.* Springer-Verlag, New York, NY, 2007.

[5] Mark A. Armstrong. *Basic Topology.* Springer-Verlag, New York, NY, 1983.

[6] Bay Area Mathematics Olympiad-12. *The Emissary* (MSRI newsletter), February 12, 2015.

[7] Ralph P. Boas Jr. Möbius Shorts. *Mathematics Magazine*, 68(2):127–127, 1995.

[8] Abraham Bosse. *Manière universelle de Mr. Desargues, pour pratiquer la perspective par petit-pied, comme le géométral.* Paris, 1648.

[9] A. Bouvier and M. George. *Dictionnaire des mathématiques.* Presses Universitaires de France, Paris, 1977.

[10] Rey Casse. *Projective Geometry: An Introduction.* Oxford University Press, Oxford, 2006.

[11] David Cox, John Little, and Donal O'Shea. *Ideals, Varieties and Algorithms: An Introduction to Computational Algebraic Geometry and Commutative Algebra.* Springer-Verlag, New York, 3rd edition, 2007. This book deeply influenced our approach to defining extended real space.

[12] H.S.M. Coxeter. *Projective Geometry.* Springer-Verlag, New York, NY, 1987.

[13] Annalisa Crannell. Writing in Mathematics; available from http://www.fandm.edu/annalisa -crannell/writing-projects-in-math-classes, 1994.

[14] Annalisa Crannell, Marc Frantz, and Fumiko Futamura. The Image of a Square. *American Mathematical Monthly*, 124(2): 99–115, 2017.

[15] Annalisa Crannell, Marc Frantz, and Fumiko Futamura. Dürer: Disguise, distance, disagreements, and diagonals! *Math Horizons*, 22(2): 8–11, 2014.

[16] Girard Desargues. Exemple de l'une des manieres universelles du S.G.D.L. touchant la pratique de la perspective sans emploier aucun tiers point, de distance ny d'autre nature, qui soit hors du champ de l'ouvrage. Paris. 1636. In *The Geometrical Work of Girard Desargues*. J. V. Field and J. J. Gray, eds. Springer-Verlag, New York, NY, 1987.

[17] R. J. Duffin. On seeing progressions of constant cross ratio. *American Mathematical Monthly*, 100(1)38–47, 1993.

[18] A. Dürer. *St. Jerome in His Study* (1514). From Wikimedia Commons, https://commons.wikimedia.org/wiki/File:Albrecht_D%C3%BCrer_-_St_Jerome_in_his_Study_-_WGA07318.jpg

[19] Howard Eves. *A Survey of Geometry (Revised Edition)*. Allyn and Bacon, Boston, MA, 1972.

[20] John Fauvel, Raymond Flood, and Robin Wilson, eds. *Möbius and His Band: Mathematics and Astronomy in Nineteenth-Century Germany*. Oxford University Press, Oxford, 1993.

[21] George K. Francis. *A Topological Picturebook*. Springer-Verlag, New York, NY, 1987.

[22] Marc Frantz. A Car Crash Solved—with a Swiss Army Knife. *Mathematics Magazine*, 84(5): 327–338, 2011.

[23] Marc Frantz. A Different Angle on Perspective. *The College Mathematics Journal*, 43(5): 354–360, 2012.

[24] Marc Frantz and Annalisa Crannell. *Viewpoints: Mathematical Perspective and Fractal Geometry in Art*. Princeton University Press, Princeton, NJ, 2011.

[25] F. Futamura, R. Lehr. A new perspective on finding the viewpoint. *Mathematics Magazine*, 90(4): 267–277, 2017.

[26] Jeremy Gray. August Ferdinand Möbius. Page 759. In *Princeton Companion to Mathematics*, Timothy Gowers, ed. Princeton University Press, Princeton, NJ, 2008.

[27] J. B. Greenough and J. H. Allen. *Allen and Greenough's New Latin Grammar*. Focus Publishing, R. Pullins & Company, Inc., Newburyport, MA, 2001.

[28] H. Brian Griffiths. *Surfaces*. Cambridge University Press, Cambridge, [U.K.]; New York, NY, 1976.

[29] Branko Grünbaum. *Configurations of Points and Lines: Volume 103 of Graduate Studies in Mathematics*. AMS, Providence, RI, 2009.

[30] Paul Halmos. How to write mathematics. *Enseign. Math.*, 16: 123–152, 1970.

[31] Nicholas J. Higham. *Handbook of Writing for the Mathematical Sciences*. SIAM, Philadelphia, PA, 1994.

[32] Merriam Webster online dictionary. https://www.merriam-webster.com/dictionary/jamb.

[33] Morris Kline. Chapter XI. Science Born of Art: Projective Geometry. In *Mathematics in Western Culture*. George Allen and Unwin Ltd., London, 1954.

[34] Steven G. Krantz. *A Primer of Mathematical Writing*. AMS, Providence, RI, 1997.

[35] Stan Lee and John Buscema. *How to Draw Comics the Marvel Way*. Simon & Schuster, New York, NY 1977.

[36] Leonardo da Vinci. *Codex Urbinas Latinus 1270 (Treatise on Painting)*. Translated by A. Philip McMahon. Princeton University Press, Princeton, NJ, 1956.

[37] MathOverflow. http://mathoverflow.net/questions/135991/visualization-of-the-real-projective-plane.

[38] Stephen Maurer. Advice for Undergraduates on Special Aspects of Writing Mathematics. *PRIMUS*, 1(1): 9–28, 1991.

[39] Martin J. Mohlenkamp. The Good Problems website. http://www.ohiouniversityfaculty.com/mohlenka/goodproblems/goodproblem.pdf.

[40] Flags problem. *The Emmissary* (MSRI newsletter), page 11, Fall 2016. https://www.msri.org/system/cms/files/275/files/original/Emissary-2016-Fall-Web.pdf.

[41] Dan Pedoe. *Geometry and the Visual Arts*. Dover, New York, NY, 2011.

[42] Anthony Phillips. (Review of) The Invention of Infinity: Mathematics and Art in the Renaissance. *AMS Notices*, 47(1):46–50, 2000. https://www.ams.org/notices/200209/rev-phillips.pdf.

[43] Clifford A. Pickover. *The Möbius Strip: Dr. August Möbius's Marvelous Band in Mathematics, Games, Literature, Art, Technology, and Cosmology*. Thunder's Mouth Press, New York, NY 2006.

[44] Piero della Francesca. *De Prospectiva pingendi*. Edizioni Ca' Foscari, 2016. http://doi.org/10.14277/978-88-6969-099-0.

[45] Donald Robertson. Desargues Theorem. https://math.osu.edu/sites/math.osu.edu/files/DesarguesTheorem.pdf.

[46] Don Row and Talmage James Reid. *Geometry, Perspective Drawing, and Mechanisms*. World Scientific, Singapore, 2012.

[47] Henry Sayre. *Writing about Art*. Prentice Hall, New Jersey, 1995.

[48] G. C. Shephard. Isomorphism invariants for projective configurations. *Canad. J. Math.*, 51(6):1277–1299, 1999.

[49] Saul Stahl. *Introduction to Topology and Geometry*. Wiley InterScience, Hoboken, NJ, 2005.

[50] N. E. Steenrod, P. R. Halmos, M. M. Schiffer, and J. A. Dieudonné. *How to Write Mathematics*. AMS, Providence, RI, 1973.

[51] Brook Taylor. *New Principles of Linear Perspective: Or the Art of Designing on a Plane the Representations of All Sorts of Objects, in a More General and Simple Method than Has Been Done Before*. London, 1719.

[52] Gerard Venema. *Exploring Advanced Euclidean Geometry with GeoGebra*. MAA, Washington, DC, 2013.

Index

Alberti, Leon Battista, 3, 155
apparent size, 19
axis, 94, 121, 123, 133, 252, 253

Bosse, Abraham, 113
box, 22, 147, 159, 178, 186
Boy's surface, 232

camera obscura, 35
car crash, 197
Cartesian coordinate system, 213
Casey angle, 205, 257
Casey's Theorem, 206, 257
cat on stairs, 20
center, 35, 79, 94, 121, 133, 150, 219, 251–254
Center-axis-point Theorem, 124, 253
Ceva's Theorem, 56, 200, 249
    converse, 59
circular product, 192, 199
collinear points, 45, 173, 205, 224, 246
concurrent lines, 45, 246
congruence, 44
cross ratio, 174, 192, 198, 256
cross-cap, 232
cube, 72, 98, 216, 254
    image of, 145, 150, 162

Desargues's Theorem, 94, 105, 113, 252
    converse, 97, 105, 252
Desargues, Girard, 3, 94, 113, 121
diagonal points, 84, 252
directed distance, 55, 174, 192, 196, 256, 257
Dürer, Albrecht, 152

elation, 121–123, 133–138, 143, 253–254
Elation Theorem, 144, 254
equivalence, 220
equivalence class, 220, 257
Eves's Theorem, 193, 257
extended Euclidean space, 68, 250
extended Euclidean plane, 66, 221, 227, 249

fence, 39, 75, 119, 178
fixed (point or line), 121, 143, 253
four color theorem, 233

GeoGebra, 96, 147, 161, 237
geometric mean, 198
golden rectangle, 185

h-expression, 192, 257
hallway, 21, 89, 242
harmonic ratio, 60, 178, 185, 249, 256
harmonic set, 60, 133, 178, 185, 207, 249, 253, 256
homogeneous line, 222, 257
homogeneous point, 220, 257
homology, 133
    harmonic, 134, 253
    period-2, 133
horizon, 27, 65, 147, 154, 183, 242

ideal
    line, 66, 68, 250
    plane, 68, 250
    point, 65, 229, 249
image, 35

images of lines, 14, 37
    perpendicular, 41
incident, 44, 73, 77, 94, 223, 246, 251, 252
Italian flag, 184

jamb, 21, 70

Klein bottle, 230

Leonardo da Vinci, 4, 18, 269
letter (three dimensional), 27, 30, 102, 119, 155, 169

Menelaus's Theorem, 59, 200, 249
mesh, 77–86, 94–97, 102, 109, 133, 251–253
    full, 86
mesh map, 79–83, 109, 111, 128, 251–253
Möbius band, 227
Möbius shorts, 229

nose, looking through, 13

object, 13, 35, 245
octahedron, 168
one-point perspective, 22, 30, 246
ordinary
    line, 66, 250
    plane, 68, 250
    point, 66, 250

Parallel Postulate, 44, 48, 248
Pedoe, Dan, 99
pencil of lines, 28, 31, 232, 246
period-2 homology, 253

perspective
  etymology, 4
  from a line, 94, 107, 252
  from a point, 94, 105, 252
perspective collineation, 81–82,
    121–128, 133–139,
    143–144, 193, 201
Perspective Proposition, 109,
    252
perspectivity, 40, 138
picture plane, 13, 245
Piero della Francesca, 99

quadrangle, 45, 84, 133, 215, 247,
    252
  complete, 84, 252
  diagonals, 46, 160, 161
  opposite sides of, 133
  sides, 46

reflection, 122, 128, 133
Roman's surface, 232

Saint Jerome in His Study,
    152
semicircle, 159, 254
shadow, 35, 93, 110, 171, 184,
    219
side view, 13, 128, 245
square, 199, 209, 215
surface, 227
  non-orientable, 230

Taylor, Brook, 35
Three points Theorem, 124,
    253
three-point perspective, 22, 31,
    246
top view, 13, 128, 245

torus, 230
translation, 121, 143
triangle, 45, 72, 84, 94, 105, 122,
    193, 247, 252
  isosceles, 55
trompe l'oeil, 152
two-point perspective, 22, 31,
    159, 246

vanishing point, 15, 215,
    245
viewer, 13, 245
viewing circle, 160, 254
viewing distance, 149, 164, 210,
    254
viewing target, 149, 164, 210,
    254

window, 4–8, 11–13